JANICE VANCLEAVE'S
Teaching the Fun of Science

John Wiley & Sons, Inc.

New York • Chichester • Weinheim • Brisbane • Singapore • Toronto

Published by John Wiley & Sons, Inc.
Published simultaneously in Canada

Design and production by Navta Associates, Inc.

All illustrations except those on pages 38, 47, 110, and 153 copyright © 2001 by Laurel Aiello.

The publisher and the author have made every reasonable effort to ensure that the experiments and activities in this book are safe when conducted as instructed but assume no responsibility for any damage caused or sustained while performing the experiments or activities in the book. Parents, guardians, and/or teachers should supervise young readers who undertake the experiments and activities in this book.

Library of Congress Cataloging-in-Publication Data
VanCleave, Janice Pratt
    Janice VanCleave's teaching the fun of science / Janice VanCleave.
        p. cm.
    Includes index
    ISBN 0-471-19163-9 (pbk.)
    1. Science—Study and teaching. I. Title: Teaching the fun of science. II. Title.

Q181 .V295 2001
507'.8—dc21

00–043857

Printed in the United States of America

10 9 8 7 6 5 4 3 2 1

Contents

Dedication

It is a pleasure to dedicate this book to Dr. Ben Doughty. Dr. Doughty is head of the department of physics at Texas A&M University–Commerce in Commerce, Texas. This special person has patiently provided me with answers to my many questions about physics and science in general. This valuable information has made this book even more understandable and fun.

Acknowledgments

I wish to express my appreciation to these science specialists for their valuable assistance by providing information or assisting me in finding it: members of the Central Texas Astronomical Society, including Johnny Barton, John W. McAnally, and Paul Derrick. Johnny is an officer of the club and has been an active amateur astronomer for more than 20 years. John is also on the staff of The Association of Lunar and Planetary Observers, where he is acting Assistant Coordinator for Transit Timings of the Jupiter Section. Paul is the author of the "Stargazer" column in the *Waco Tribune-Herald.* Dr. Glenn S. Orton, a Senior Research Scientist at the Jet Propulsion Laboratory of California Institute of Technology: Glenn is an astronomer and space scientist who specializes in investigating the structure and composition of planetary atmospheres. He is best known for his research on Jupiter and Saturn. I have enjoyed exchanging ideas with Glenn about experiments for modeling astronomy experiments.

A special note of gratitude to these future educators who assisted by pretesting activities and/or by providing scientific information—the elementary education students of Dr. Belinda Anderson, Dean of the School of Education, Lambuth University, Jackson, Tennessee: Michele Bowen, Brooke Newsom, Christy Crawford, Alison Holt, Starr Chestosky, Regina Dorris, Janet Robinson, Paul Mayer, Sara Hatch, Marcia Law, Crystie Sikes.

These very special people provided not only encouragement but invaluable scientific information. Virginia Malone, science assessment consultant; Sally A. DeRoo, science educator at Wayne State University, Detroit, Michigan; Holly Harris, science educator at China Spring Intermediate, China Spring, Texas; Robert Fanick, a chemist at Southwest Research Institute in San Antonio, Texas; and Anne Skrabanek, homeschooling consultant, Perry, Texas.

From the Author

Kids love science. They get excited about attracting objects with a magnet, observing the metamorphosis of a butterfly, growing crystals, and finding patterns in the stars. In other words, they enjoy investigating topics dealing with physical science, life science, and Earth and space sciences. With your guidance, this natural curiosity can be encouraged, resulting in a deeper understanding of these disciplines. The investigations in this book will help you direct your students toward developing a basic knowledge of these sciences, and this knowledge will in turn provide a foundation on which to build more scientific knowledge and a love of scientific exploration.

The prime goal of this book is to provide opportunities in which to engage students in investigations. This hands-on approach encourages students to understand science concepts, gives them ways to apply the concepts, and introduces and reinforces the skills they need to become independent investigators.

The basic outline and objectives of each section of the book follow the standards for kids aged 8 to 12 set forth in the *National Science Education Standards.* Although all the standards are not addressed in this book, many basic benchmarks for this age group are represented in the investigations.

The investigations in this book can be adapted for students of different grade levels by increasing or decreasing the amount of information provided. For example, in investigation 1, "Different Kinds," students model molecules composed of two atoms, but you can also introduce older students to bond angles and show how molecules, such as water with two hydrogen atoms and one oxygen atom, are combined.

Assessment of individual student achievement is ongoing, but teachers need tangible materials on which to base a student's grade. For the investigations in this book, I've suggested multiple methods of assessment, such as investigative reports (from pictorial work for younger students to written reports for older students), models, and other creative expressions of understanding, as well as traditional paper-and-pencil tests. My advice to teachers is not to let the need for evaluation of student work take the fun out of discovering the wonders of science. Try to balance the free spirit of discovery with the business of recording and sharing scientific data.

Guidelines for Using Science Investigations Successfully in the Classroom

Review the Teaching Tips

The book is organized into four sections, one addressing the scientific method and three the branches of science including physical, life, and Earth and space. Each of the latter three sections and the subsections within each open with a brief introduction identifying general objectives for the section. Together these three sections comprise a total of 75 investigations for students.

Each investigation is accompanied by an overview giving tips for teaching the investigation, including the science benchmarks addressed by the investigation, expectations of student learning, suggestions for preparing materials, a miniglossary of new terms to which students will be introduced, background information and interesting facts on the topics investigated, and one or more extensions.

Terms used in the investigation are **boldfaced** and defined in the investigation, the teacher tips, and the glossary in the back of the book. These terms as well as other key science terms appear in *italics* in the section introductions. Key science terms in the teaching tips, other than those boldfaced, appear in italics. You may wish to enrich and expand the science content of your lessons by introducing the key terms to your students. All key terms are listed in the index, with reference to other parts of the book where the terms are discussed.

Measurements used in the procedures of the investigations, as well as in the overviews, are given in English units with approximate metric equivalents in parentheses. This allows the reader to use either the English or metric system, but is not intended to reflect precise equivalencies between the two systems.

Get to Know the Investigations

Read each investigation completely before starting, and practice doing the investigation prior to class time. This increases your understanding of the topic and makes you more familiar with the procedure and the materials. If you know the investigation well, it will be easier for you to give instructions, answer questions, and expound on the topic.

Investigations follow a general format:

1. **Purpose:** The goal of the investigation.
2. **Materials:** A list of necessary supplies (common household items) needed for each student or group.
3. **Procedure:** Step-by-step instructions.
4. **Results:** For some investigations, a data table is provided for students to record their observations. In other investigations, an explanation stating exactly what is expected to happen is given. This is an immediate learning tool. If the expected results are achieved, your students have immediate positive reinforcement. If the results are not the same, encourage them not to change their data. Point out that scientists may not achieve expected results, but they always accurately record the results observed. To encourage this, you might devise a rubric system, an evaluation that rewards students for successfully completing the investigation rather than for the correctness of the results.
5. **Why?:** Investigations explain why the results were achieved, in terms students will understand. Students are introduced to new science terms as they work their way through each investigation. These new terms appear in **boldface** and are defined in the "Why?" section of the investigation.

All the terms from the 75 investigations are included in the glossary at the back of the book.

Collect and Organize Supplies Well Ahead of Time

You will be less frustrated and more successful if you have all the necessary materials for the science investigations ready for instant use. Decide whether the students will be doing the science investigation individually or in groups, and calculate from that how much of each material you will need for the class. I prefer to designate a place in the classroom where the supplies will be placed each time a science discovery time is scheduled. I separate the materials and put each type of material in its own box or area of the table. I also provide boxes or trays for the students to use to carry the materials to their work area. You may want to have your students help gather and organize supplies.

Set Up Collaborative Groupings and Assign Jobs

Forming students into teams to conduct the science investigations helps you manage the class and provides the best opportunity for them to learn not just the science but also how to work together. Groups of four are ideal, but smaller or larger groups are also acceptable, and may in some cases be preferable. Each group works as a team to collect and analyze data, but reports may be group or individual efforts. Not only does collaboration enhance student learning, but working in groups also reduces the number of supplies needed.

Assign each group member a job or allow the group to decide who does which job. This will make science time both a fun-filled adventure for the students and a time that you look forward to as one of the easiest and most organized periods of the day. Having students wear lab aprons keeps them neat and helps get kids excited about acting as scientists. Aprons can also provide immediate identification of each student's job if you give out a different-colored apron for each job.

Suggested Job Titles and Duties

DIRECTOR This team member leads the science investigation. The director is the facilitator, but each child should do part of the science investigation. The director determines what part of the investigation each group member performs. The director can also be the one to report problems to you that the group might be having. One way the director can notify you of the group's progress is to use three colored cups stacked on top of each other as signals. The top color indicates the need of the group: red (need help immediately), yellow (we have a question when you have time), green (all is well).

SUPPLY MANAGER This team member will pick up needed supplies for the science investigation from the supply table and return any unused supplies to the table at the end of the science investigation. Each supply manager will need a copy of the materials list for the investigation. It helps to have all the supply managers assemble in front of the supply table at the same time so that you can identify the materials to them and give any special instructions for transporting and using the materials. The supply manager and the waste manager might be the only students allowed to move around the room.

RECORDER This team member records the observations made by the group. This can be in the form of drawings and/or written data, such as tables. The recorder collects and hands in any papers that are to be turned in by the group. (For some investigations, individual record keeping is required, so a group recorder is not needed.)

WASTE MANAGER This team member is responsible for discarding all used materials in their proper place. The waste manager should also make sure that the work area is clean and ready for the next classroom activity. The waste manager could also be the timekeeper. It is important to complete the investigation so there is ample time for cleanup.

Supervise the Science Investigation

Instruct your students to read through each science investigation before beginning and to follow each step very carefully, never skipping or adding steps. Emphasize that safety is of the utmost importance and that the instructions should be followed exactly. To ensure that the students understand the procedure, you may want to demonstrate all or part of the procedure before they start on their own. You might stop short of showing the final step so that the students experience seeing the results for the first time themselves.

Help Students Analyze the Results

As noted previously, it is best if you perform the science investigation yourself in advance so you know what to expect. Then if the students' results are not the same as those described in the science investigation, you will be better prepared to help them figure out what might have gone wrong. First, go over the procedure, step-by-step, with the individual or group to make sure that no steps were left out. If all the steps were completed, try asking the students leading questions. The students can then provide their hypotheses as to why the results were not achieved. Analyze the materials. I like to ask questions such as "Do you think our tap water has too many chemicals in it?" or "Would a different brand of gelatin affect the results?"

Next, analyze other variables in the classroom environment, such as lighting, temperature, and humidity. Introduce questions such as "Did turning the lights off at night affect the results?" or "Do you think the excess moisture in the air affected the results?" While I prefer to brainstorm with students, sometimes I just have the group reread the instructions and redo the science investigation. This provides an opportunity to point out to students that scientists always redo experiments to confirm their results.

When the results do work out, refer to the "Teaching Tips" and "Why?" sections in the investigation to help you provide the scientific explanations.

Encourage Students to Report on Their Results

Now's the time to point out that good science is not just careful investigation, it's also accurate documentation. You can assign the students individual reports, or a group report to be written by the recorder with input from the whole group. Reports can range from simple drawings representing the results of the investigation to written reports that summarize the procedure and data and give a conclusion that explains why the results were achieved. Combined class results can be written on a chalkboard and used for class discussion. Having students do research on the experiment topic helps them learn valuable research skills.

Make Suggestions for Further Investigations

After the investigation has been performed, direct the groups in thinking about how the results would be affected if one part of the experiment were changed. For example, if a specific length and kind of string are used, guide them in examining the possibility of trying the investigation with different lengths or kinds of string (but be sure to change only one variable at a time!). Start this type of thinking with a question such as "I wonder what would happen if you used a longer string?" Solicit hypotheses from each group and select those that are most practical for further investigations. The procedure for the original investigation may only need some changes to test the hypothesis. Data will be collected and the conclusion will indicate whether the results support the hypothesis. Point out that the results only do or do not support the hypothesis and are never considered as indicating that the hypothesis is right or wrong. For specific ideas, see the Extension section in the Teaching Tips of each chapter.

Science as Inquiry

The tool for the science inquiry approach is the *scientific method*. This method is the process of identifying a problem, thinking through the possible solutions to the problem, and testing each possibility for the best solution. The scientific method involves the following: *research* (the process of collecting data about a topic being studied), a *problem* (a scientific question to be solved), a *hypothesis* (an idea about the solution to a problem, based on knowledge and research), an *experiment* (the process of testing a hypothesis or answering a scientific question), and a *conclusion* (a summary of the results of an experiment and how the results relate to the hypothesis or how it answers the problem question).

Although these steps of the scientific method are named in a specific order, scientists do not always follow this order. Research is named as the first step in the scientific method, but research is an ongoing part of any investigation. Not all steps of the scientific method are part of every classroom investigation. For example, many classroom investigations have a problem, but do not require a written hypothesis. Even so, ideas about the answer to the problem generally come to mind. Some investigations do not have an experiment per se. For example, the behavior of classroom animals can be observed and conclusions made from the data collected.

Classroom science investigations are designed to help students develop six skills: (1) asking questions (or posing a problem); (2) predicting what they expect to observe (or forming a hypothesis); (3) planning and conducting investigations (including experiments to test their hypothesis); (4) collecting observations (data); (5) organizing, examining, and evaluating data by constructing tables, graphs, charts, and maps; and (6) drawing conclusions by comparing their hypothesis (expected observations) with their data (actual observations). If a hypothesis is not required, the conclusion would be a summary of the results, including the answer to the question asked.

Physical Science

Physical science includes chemistry and physics. *Chemistry* is the study of the way materials are put together and how they change under different conditions. *Physics* is the study of energy and matter and the relationship between them. Kids enjoy learning about physical science because it deals with things that they like to do, such as making slime and experimenting with magnets. In physical science, it's most important to learn properties of matter and the changes they undergo, as well as the meaning of energy and its transfer, including motions and forces of objects.

A

Properties and Changes of Properties in Matter

Everything in the universe is composed of matter. Atoms are the basic building blocks of matter. All atoms today have been around since the beginning of the universe. Atoms combine to form new substances, then break apart and recombine in different ways over and over again. The atoms in your body might have been in the body of a dinosaur millions of years ago. Mostly, atoms change only by losing or gaining outer parts called *electrons*. In this section, students will use models to discover the differences between the different building blocks of matter: atoms, elements, and compounds.

Matter exists in different forms called *phases,* and each phase has its own *physical properties* (color, shape, weight, etc.). In this section, students will learn to identify the different phases of matter (solid, liquid, gas) and will discover that matter has physical properties by observing and using appropriate tools to identify specific physical properties.

Substances in different phases of matter can be mixed in different ways. When substances are combined to form a mixture, they are easily separated. An example is salt and water, which can be separated by evaporating the water. But if the substances chemically combine to form a compound, they are not easily separated. An example is the chemical combination of sodium and chlorine, which produces sodium chloride (table salt). It usually requires energy for a chemical combination to occur. In this section, students will demonstrate that some combinations produce a mixture that maintains the physical properties of its ingredients while other combinations do not.

Teaching Tips for the Investigation

Different Kinds

Benchmarks

By the end of grade 5, students should know that

- Materials may be composed of parts that are too small to be seen without magnification.
- When a new material is made by combining two or more materials, it has properties that are different from those of the original material.

By the end of grade 8, students should know that

- Materials made of different parts are called *systems*.
- Matter is made of atoms.
- Atoms of any one element are alike, but are different from atoms of another element.

In this investigation, students are expected to

- Make models of systems and their separate parts and models of atoms (parts) and molecules (systems).
- Understand that parts combine to form systems.
- Distinguish between atoms, molecules, elements, and compounds.

Preparing for the Investigation

Prepare a Matter Data table and make one copy for each student. Prepare a resealable plastic bag of small and large, different-colored gumdrops for each student or group. Number each bag and write on it "Do Not Eat." Place 9 to 12 different-colored gumdrops in each bag. Make an effort to have a different number of each color gumdrop in the bags. Have students write their bag number on their copy of the Matter Data table.

Presenting the Investigation

1. Introduce the new science terms:

   atom The smallest unit of an element; a building block of matter.

   bond A force that links atoms together.

   compound A substance made of molecules that are alike.

   diatomic molecule A molecule made up of two atoms of the same kind.

   element A substance made up of atoms that are alike.

   formula A symbolic representation of a molecule.

   mass An amount of material.

   matter Anything that occupies space and has mass.

   molecule A group of two or more atoms held together by bonds.

   substance A material made of one kind of matter.

2. Explore the new science terms:

   - Matter is the stuff the universe is made of.
   - Most elements exist as single atoms, but some exist as larger units called molecules, such as diatomic molecules. For example, the symbol for one atom of hydrogen is H, but hydrogen atoms are rarely alone. It is a diatomic molecule (H_2).
   - Examples of compounds are water (H_2O) and table salt, sodium chloride (NaCl).
   - Atoms and molecules are in constant motion.
   - Models of molecules indicate the kind and number of atoms and their geometric spacing.
   - Chemical symbols for elements consist of one or two letters. If the symbol consists of one letter, it is capitalized, such as C for the element carbon. If the symbol consists of two letters, the first letter is capitalized and the second is lowercased, such as Co for cobalt. Symbols are always written in block letters, never in cursive.
   - A formula represents two or more atoms, such as the diatomic element, H_2, and the compound water, H_2O.

Did You Know?

No matter how substances within a closed system interact with each other, the number of atoms remains the same. This law is called the *conservation of matter.*

EXTENSION

Systems are a combination of parts forming a whole unit. Systems may combine to form larger systems. How is an atom a system and also part of a system? (An atom is made up of parts: a center called a *nucleus,* which contains *protons* (positively charged particles) and *neutrons* (neutrally charged particles). Outside the atom are *electrons* (negatively charged particles). Atoms combine to form molecules, molecules combine to form parts of a living organism, living organisms combine to form populations, and so on.

Different Kinds

PURPOSE

To make models of atoms and molecules.

Materials

12 or more gumdrops of various sizes and colors
Matter Data table
crayons
pen
6 or more toothpicks

Procedure

1. Look at the gumdrops. Each gumdrop represents one atom. Each color and size of gumdrop represents a different kind of atom. In the Matter Data table under "Atoms" in the Substance column, make a colored drawing like the one shown of each kind of gumdrop.

2. Group the gumdrops by color and size. In the "Symbol/Formula" column of the Matter Data table, write a symbol to indicate each kind of atom represented. For example, the symbols could be "Lr" for a large red gumdrop and "Sr" for a small red gumdrop.

3. Stick a gumdrop on each end of a toothpick. The joined gumdrops represent a molecule. Continue using pairs of gumdrops and a toothpick to model as many different kinds of molecules as possible.

4. Group together the molecules that are alike.

5. Under "Molecules" in the "Substance" column of the table, make a colored drawing like the one shown of each kind of molecule.

6. In the "Symbol/Formula" column of the table, write a chemical formula for each different kind of molecule. For example, if a molecule is made up of two gumdrops, one Lr and one Sr, the formula might be LrSr. (Note that there is no space between the symbols.) If a molecule is made up of two gumdrops of the same color, one Lr and one Lr, the formula would be Lr_2. (Note that the number is a subscript.)

Results

You have filled in each column of the Matter Data table. The kind and number of atoms and molecules varies depending on the color and size of the gumdrops and how they are combined.

Why?

Matter is anything that occupies space and has **mass** (an amount of material). A **substance** (material made of one kind of matter) made of two or more atoms that are alike is an **element.** The smallest unit of an element is an atom. The individual gumdrops represent atoms, and each kind of gumdrop "atom" differs from other kinds by its size and/or color. For example, a large red gumdrop atom is different from a small red one, and a small green gumdrop atom is different from a small or large red one.

Gumdrop **molecules** (a group of two or more atoms held together by bonds) are made up of gumdrop atoms linked by toothpicks that represent **bonds** (forces that link atoms together). A **compound** is a substance made of molecules that are alike. Molecules made up of two atoms of the same kind are called **diatomic molecules,** as represented by two like gumdrops linked by a toothpick. A possible **formula** (a symbolic representation of a molecule) for a diatomic molecule would be Lr_2.

MATTER DATA TABLE	
Substance	**Symbol/Formula**
Atoms	Lr
	Sr
Molecule	Lr_2

Curved

Benchmarks

By the end of grade 5, students should be able to
- Measure liquid materials in prescribed amounts.

By the end of grade 8, students should be able to
- Choose appropriate units for reporting measurements.

In this investigation, students are expected to:
- Use a model of a graduated cylinder to measure the physical property of volume in SI units.

Preparing for the Investigation

See appendix 1 for a pattern and instructions for making a model of a graduated cylinder. Depending on how much time you have, you can either make one model for each student or group in advance, or you can have the students make the models as part of the investigation. If you wish to have reusable models, laminate the sheets before cutting.

Presenting the Investigation

1. Introduce the new science terms:

   graduated cylinder An instrument used to measure volume.

   International System of Units (SI) The measurement system used primarily in science and technology. Commonly called the **metric system.**

   liter (L) The SI or metric unit of volume.

   meniscus The curved upper surface of a column of liquid.

   metric system See **International System of Units (SI).**

   milliliter (ml) One-thousandth of a liter.

   volume The amount of space something occupies.

2. Explore the new science terms:
   - *Physical properties,* such as mass and volume, are characteristics of matter that can be measured and observed without changing the makeup of the substance. Equal volumes of different substances usually have different masses. (*Density* is a ratio of mass to volume. For more information about density, see investigation 64.)

- Things that are not matter include visible light and sound waves.
- In a given place, equal masses experience equal gravitational force, which is called *weight*. So on Earth, if the mass is known, the weight can be determined. Basically 1 pound of weight equals 454 g of mass.
- For everyday purposes, there are two basic measuring systems: (1) imperial or English and (2) metric. The imperial system was developed in the United Kingdom and is commonly used in the United States. The metric system was created by French scientists in the late eighteenth century and is used in most parts of the world. The metric system is a convenient system because units of different sizes are related by powers of 10. The metric system was used by scientists until 1960, when a revised system based on the metric system was established. This new system is called the International System of Units (SI).
- All liquids in a container have a meniscus, but if the container is broad, such as a drinking glass or a jar, the curve is not always noticeable.

Did You Know?

While the surface of most liquids curves downward, the surface of mercury curves upward. *Caution: Do not allow students to handle mercury, because it is poisonous.*

E X T E N S I O N

Encourage the class to work independently with the models. Observe the progress of students individually to make sure they can find the numbers on the model. Write more volumes on the chalkboard that can be demonstrated by the model. The model can be used to evaluate student understanding. Do this by giving each student a volume to represent with the model. Watch as the student moves the strip to place the center of the meniscus on the mark for the given volume.

Curved

PURPOSE

To read a graduated cylinder.

Materials

model of a graduated cylinder (provided by teacher, or see instructions on separate sheet)

Procedure

1. Look at the curved end of the liquid strip in the model. Move the strip so that the center of the curve is on 9.

2. Move the strip so that the center of the curve is on 6.5 (the mark between 6 and 7).

Results

Volumes of 9 ml and 6.5 ml are found on the model graduated cylinder.

Why?

One way of measuring matter is to find its **volume,** the amount of space it occupies. In the **International System of Units (SI),** commonly called the **metric system,** liquid volume is measured in **liters.** A **milliliter (ml)** is one-thousandth of a liter.

Liquid volume can be measured using a container called a **graduated cylinder.** The liquid strip in the model graduated cylinder represents a column of liquid. The curve is called a **meniscus** and represents the curved surface of a liquid.

Each numbered division on the model is equal to 1 ml. The marks halfway between the numbered divisions equal one-half of a milliliter, or 0.5 ml. So the first reading of 9 represents that the graduated cylinder contains 9 ml of liquid. The second reading of 6.5 represents that the graduated cylinder contains 6.5 ml of liquid.

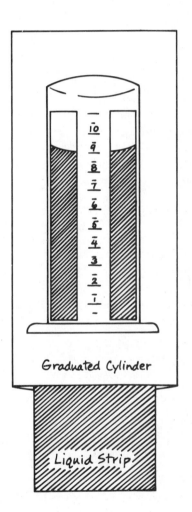

Graduated Cylinder

Liquid Strip

Phases

Benchmarks

By the end of grade 5, students should know that

- Water can be a solid, a liquid, or a gas and can change back and forth from one form to another.

By the end of grade 8, students should know that

- The phase of matter of a substance depends on the motion of its atoms. Atoms of solids have the least motion; atoms of liquids have moderate motion; atoms of gases have the most motion.

In this investigation, students are expected to

- Distinguish between the physical properties of the three phases of matter.
- Identify matter as liquids, solids, and gases.
- Understand that a liquid changes to a gas by evaporation.

Preparing for the Investigation

The perfume used in the investigation can be any inexpensive brand, an aromatic oil, or a food flavoring, such as vanilla. Point out that the perfume is used because its smell assists in detecting the presence of the invisible gas formed when the perfume evaporates.

Presenting the Investigation

1. Introduce the new science terms:

    diffuse To spread freely and become evenly distributed.

    evaporate To change from a liquid to a gas.

    fluid A material that flows: gas or liquid.

    gas A substance in a phase of matter characterized by no definite shape or volume.

    liquid A substance in a phase of matter characterized by a definite volume but no definite shape.

    phases of matter The forms in which matter exists. The three major phases of matter are solid, liquid, and gas.

    physical properties Characteristics of matter that can be measured and observed without changing the makeup of the substance.

    physical reaction A change in which no new substances form.

    solid A substance in a phase of matter characterized by a definite shape and volume.

2. Explore the new science terms:

    - The phase of matter of a substance depends on the motion of its atoms. The atoms in solids are closely locked in position and can only vibrate. The atoms in liquids are more loosely bound together and can slide past each other. The atoms in gases are not connected, but they do occasionally collide.

    - Physical properties of matter include phase, size, color, taste, and melting and boiling points.

    - Evaporation is an example of a physical reaction, because even though the substance changes from a liquid to a gas, it is still the same substance.

    - A liquid evaporates when the faster-moving molecules at its surface have enough energy to break away from each other.

    - When a liquid evaporates, individual molecules of that substance leave the surface of the liquid and mix with the air above the liquid's surface.

    - You know when something smelly, such as cabbage, is cooking without going into the kitchen. This is because when cabbage is cooked, particles carrying the odor diffuse or spread freely throughout the house.

Did You Know?

When you go swimming or take a bath, you do not feel wet while you are in the water. It is only after you get out of the water that you have the sensation of being wet. This wet feeling is actually a response to the rapid cooling of your skin. When you come out of the water into the air, the water on your skin evaporates, taking heat away from your body.

EXTENSION

Ask the students for examples of each phase of matter from daily life. Here are some examples: solid—sugar, clothes, furniture; liquid—water, sodas, milk; gas—air, propane (used in camping stoves), carbon dioxide (exhaled gas).

Phases

3

PURPOSE

To observe the physical properties of the phases of matter.

Materials

marker
three 3-ounce (90-ml) paper cups
tap water
ice cube
perfume
3 index cards
3 small paper or plastic bowls

Procedure

1. Use the marker to label one of the cups "Liquid," the second cup "Solid," and the third cup "Gas."
2. Fill the Liquid cup about half full with water. Cover it with an index card.
3. Place the ice cube in the Solid cup. Cover it with an index card.
4. Put a drop of perfume in the Gas cup and cover it with the remaining index card.
5. Place all three cups together on a table.
6. Remove the index card and observe the contents of the Liquid cup. Then pour the contents of the cup into one of the bowls, and observe the shape of the contents of the bowl. Note any odor that may result from pouring the cup's contents into the bowl.
7. Repeat step 6 first with the Solid cup, and then with the Gas cup, using a separate bowl for each cup.

Results

The ice stays the same shape when poured into the bowl, but the water spreads out. The water and ice have no apparent smell, but even though the perfume seems to have disappeared, the smell of the perfume can be detected.

Why?

Physical properties are characteristics of matter that can be measured and observed. **Phases of matter** are the forms in which matter exists. The three major phases of matter are solid, liquid, and gas. Phases of matter are physical properties of matter. **Solids** have a definite shape and volume. The ice was the same shape and volume in the bowl as in the cup. **Liquids** have a definite volume but no definite shape. The amount of water was the same in the bowl as in the cup, but the liquid water spread out in the bowl. **Gases** have no definite shape or volume. Materials that flow, such as gases and liquids, are called **fluids.** The liquid perfume in the cup **evaporated** (changed from a liquid to a gas), which is an example of a **physical reaction** (a change in which no new substances form). The gas formed by the evaporation of the perfume was made up of individual compound units of the perfume. These units of perfume **diffused** (spread freely) throughout the air in the room and reached your nose.

4 Teaching Tips for the Investigation

Slime

Benchmarks

By the end of grade 5, students should know that

- When a new material is made by combining two or more materials, it has properties that are different from those of the original materials.

By the end of grade 8, students should know that

- No matter how substances within a closed system interact with each other, the number of atoms remains the same. This law is called the *conservation of matter.*

In this investigation, students are expected to

- Demonstrate a chemical reaction.
- Identify a cross-linked polymer by its physical properties.
- Identify the chemical properties of a substance.

Preparing for the Investigation

While the materials used in the investigation are non-toxic, make sure the students do not eat the slime. To prevent the material from being used improperly, collect all the slime in a large resealable plastic bag at the end of the experiment, then either discard the bag of slime or refrigerate it for an extension. (The slime may be kept in the refrigerator for 2 months or more in the resealable bag.)

Presenting the Investigation

1. Introduce the new science terms:

    chemical properties Characteristics that describe the behavior of a substance when its identity is changed.

    chemical reaction A process by which atoms interact to form one or more new substances.

    cross-link A chemical bridge between polymer molecules.

    polymer A very long chainlike molecule.

2. Explore the new science terms:

    - The ability of a substance to burn is a chemical property. The process of burning is a chemical

reaction. Generally a chemical reaction cannot be reversed. You cannot unburn the paper.

- A polymer is made by bonding (linking together) many small molecules called *monomers* (one unit).
- Polymers, such as concrete, glass, plastic, and rubber, are made up of long chains of very large molecules called *macromolecules.*
- Liquid school glue contains polymers that are tangled together like a bowl of spaghetti. When starch is added to the glue, a chemical reaction occurs. The starch forms cross-links between polymer molecules much like rungs link the two sides of a ladder.

Did You Know?

In 1943, James Wright, a scientist for General Electric, was looking for an inexpensive substitute for rubber. Wright accidentally made a slime that stretched and bounced, but was not useful as a rubber substitute or for anything else. Peter Hodgson Sr. saw the slime at a party and bought the rights for it from GE. In 1950, Hodgson packaged the slime in a plastic egg and sold it as a toy. It was called "Silly Putty."

EXTENSION

Discover the effect of temperature on the slime. Place samples in resealable bags in a freezer or in a cooler with ice for 1 or more hours. Other samples in resealable bags can be placed in a warm area, such as in direct sunlight. You may wish to heat a sample in a microwave for 1 minute. *Caution: The hot slime can burn your skin. Allow it to cool before touching.* Generally the change in temperature does not affect the properties of the slime once it returns to room temperature, as long as the slime is not dehydrated. The author tested the slime at the geographic south pole at a temperature of –20°F (–28.9°C). The slime froze solid in a short period when outdoors, but retained its usually slimy properties once it defrosted when indoors.

Slime

PURPOSE

To demonstrate a chemical reaction.

Materials

spoon
1 teaspoon (5 ml) liquid starch
1 teaspoon (5 ml) white school glue
food coloring
12-inch (30-cm) -square sheet of waxed paper
timer

Procedure

1. Using the spoon, mix the starch, glue, and a drop of food coloring in the center of the sheet of waxed paper. Continue to stir the materials until they form a substance that begins to separate from the waxed paper.

2. Allow the substance to stand on the waxed paper for 3 to 4 minutes. Then roll the substance into a ball with your fingers, and knead it with your hands for about 1 minute. You have made slime.

3. Try these experiments:
   - Roll the slime into a ball and drop it on a smooth surface.
   - Set the ball of slime on a table and observe it for about 30 seconds.
   - Hold the slime in your hands and *quickly* pull the ends in opposite directions.
   - Hold the slime in your hands and *slowly* pull the ends in opposite directions.

Results

You have made a soft, pliable material that bounces slightly when dropped, spreads out when not confined, breaks apart if pulled quickly, and stretches if pulled slowly.

Why?

When you combine certain materials, their molecules do not simply mix but interact and undergo a **chemical reaction.** This means that a new substance unlike any of the substances that went into it is formed. In this investigation, the substance formed is called slime, a cross-linked polymer. A **polymer** is a very long chainlike molecule. **Cross-links** are chemical bridges between the polymer molecules. Slime is an unusual substance in that when pressure is quickly applied, it breaks like a solid. When left alone, it slowly flows like a liquid to take the shape of whatever container it is in. This behavior of slime describes its physical properties. But the behavior of starch when mixed with glue is an example of a **chemical property** (describes how one substance reacts with another) of starch, which is that it forms cross-links with the polymers in glue.

Stretchy

Benchmarks

By the end of grade 5, students should know that

- Things can be done to change the properties of a material, but all materials do not change the same way.

By the end of grade 8, students should know that

- For the results of an experiment to be clearly attributed to one variable, only one variable at a time must be changed.

In this investigation, students are expected to

- Identify the physical property of elasticity.
- Compare the elasticity of different materials.

Preparing for the Investigation

Prepare an Elastic Data table and make one copy for each student.

Presenting the Investigation

1. Introduce the new science terms:

   contract To draw together.

   elasticity The physical property of being able to return to the original length or shape after being stretched.

   standard A material to which other materials are compared.

2. Explore the new science terms:
   - Rubber bands and gummi worms are both elastic. Ask your students to think of things that are elastic, such as a basketball or a bungee cord. When stretched out of shape, elastic materials contract, or draw back to their original shape.

- To determine how elastic a material is, you have to have something to compare it to. This would be something commonly known, called a standard. In this investigation, the elasticity of the gummi worm is compared to the standard of the elasticity of the rubber band.

Did You Know?

Rubber originally meant a natural, elastic product obtained from the secretion of certain plants. Today the term is applied to a class of materials having the unique property of high elasticity. A strip of rubber can be stretched to several times its original length without breaking, and will return instantly to that length when released.

EXTENSIONS

1. Investigate the effect of temperature on the elasticity of a gummi worm. Do this by first cooling the gummi worms in a refrigerator. If a refrigerator is not available, place the gummi worms in a resealable plastic bag. Then place the bag of gummi worms in a larger resealable plastic bag containing ice. Next, heat the gummi worms by placing them in a sunny area.

2. Determine if the elastic properties of a gummi worm change with repeated stretching. Repeat the investigation several times, using the same gummi worm.

Stretchy

PURPOSE

To determine how elastic a gummi worm is.

Materials

rubber band
Elasticity Data table
scissors
gummi worm
metric ruler

Procedure

1. Without stretching the rubber band, cut a section that is the same length as the gummi worm.

2. Place the gummi worm along the edge of the ruler. Measure the worm to the nearest millimeter. In the Elasticity Data table, record this as the Length at Start.

3. Stretch the gummi worm as far as possible without breaking it, and record the greatest length as the Length When Stretched.

4. Release the gummi worm, wait for it to stop drawing together, and again measure its length. Record this as the Length at End.

5. Repeat steps 2 to 4 with the piece of rubber band.

Results

The gummi worm is found to be either more, less, or equally elastic as the rubber band, depending on the data recorded. The author found the worm to be slightly less elastic, about fifteen-sixteenths as elastic as the rubber band.

Why?

Elasticity is the ability of a material to return to its original length or shape after being stretched. A rubber band is generally considered to be perfectly elastic, meaning it returns to its original length after being stretched. The rubber band is used as the **standard** (a material to which other materials are compared) against which you are measuring the elasticity of the gummi worm. A gummi worm will generally **contract** (draw together) to almost the same length it started at. So the elasticity of the gummi worm is great, but the author's worm was slightly less than that of the rubber band.

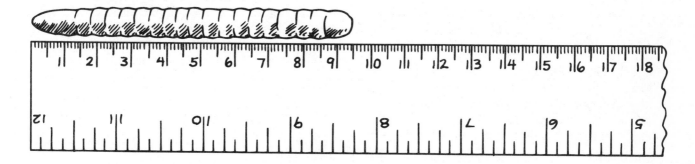

ELASTICITY DATA			
Test Material	Length at Start	Length When Stretched	Length at End
gummi worm			
rubber band			

Mixtures

Benchmarks

By the end of grade 5, students should know that

- Materials can be described in terms of what they are made of and their physical properties.

By the end of grade 8, students should know that

- A system may be thought of as containing subsystems and as being a subsystem of a larger system.

In this investigation, students are expected to

- Distinguish between homogeneous and heterogeneous mixtures.
- Identify changes that can occur in the physical properties of the ingredients of solutions, such as salt dissolving in water.

Preparing for the Investigation

Remove the labels from the water bottles. The size of bottle is not critical, but 20-ounce (600-ml) bottles work well. Stick-on labels can be used instead of masking tape.

Presenting the Investigation

1. Introduce the new science terms:

   dissolve To break up and thoroughly mix with another substance, as salt in water.

   heterogeneous Not visibly the same throughout.

   homogeneous Visibly the same throughout.

   mixture A combination of two or more substances. Mixtures may be heterogeneous or homogeneous.

   solution A homogeneous mixture in which one substance is dissolved in another.

2. Explore the new science terms:

   - Matter can be divided into two basic groups, homogeneous materials and heterogeneous materials. Each of these groups can be divided further as shown in the diagram.
   - A *substance* is a homogeneous material consisting of one particular kind of matter—element or compound.
   - The substances that make up a mixture can be separated. A homogeneous mixture has parts too small to be seen that do not separate upon standing but can be separated by other means. For example, salt and water can be separated by allowing the water to

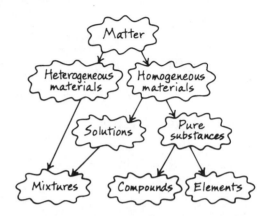

evaporate, leaving the dry salt. A heterogeneous mixture has parts that are easily seen and often separate upon standing or by other means. For example, iron filings and salt can be separated by removing the iron filings with a magnet.

- Solutions occur when one substance dissolves in another. The substance that dissolves, called a *solute*, breaks into small particles and diffuses throughout the substance it is dissolving in, which is called a *solvent*.
- Heterogeneous mixtures have two or more distinctly different parts that are easily observed, such as a mixture in a fruit salad.

EXTENSION

Explain that solutions can be mixtures of solids dissolved in solids, liquids, or gases; liquids dissolved in solids, liquids, or gases; and gases dissolved in solids, liquids, or gases. Note that a solution can be in any phase of matter and that the phase of the solution is the same as the phase of the solvent. Ask students to research the different types of solutions and make a table similar to the Types of Solutions table, showing an example of each type.

TYPES OF SOLUTIONS			
Solute	Solvent	Solution	Example
solid	solid	solid	copper and zinc (brass)
solid	liquid	liquid	salt in water (seawater or blood)
solid	gas	gas	carbon particle in air (soot in air)
liquid	solid	solid	mercury in silver (dental filling)
liquid	liquid	liquid	chocolate syrup in milk (chocolate milk)
liquid	gas	gas	water in air (humid air)
gas	solid	solid	poisonous gases on carbon (poisons trapped in charcoal gas mask filter)
gas	liquid	liquid	carbon dioxide in water (soda)
gas	gas	gas	nitrogen and oxygen (air)

Mixtures

PURPOSE

To determine the basic difference between homogeneous and heterogeneous mixtures.

Materials

2 empty 20-ounce (600-ml) water bottles with caps
tap water
½ teaspoon (2.5 ml) table salt
masking tape
pen
timer
¼ cup (63 ml) liquid cooking oil

Procedure

1. Fill one of the bottles about three-fourths full with water.
2. Add the salt to the bottle of water and secure the cap.
3. Use the tape and the pen to label the bottle "Homogeneous Mixture."
4. Shake the bottle vigorously 20 or more times. Allow the bottle to stand for 2 to 3 minutes.
5. Fill the second bottle about half full with water.
6. Add the cooking oil to the bottle and secure the cap.
7. Use the tape and the pen to label the bottle "Heterogeneous Mixture."
8. Shake the bottle vigorously 20 or more times. Allow the bottle to stand for 2 to 3 minutes.
9. Compare the contents of the bottles.

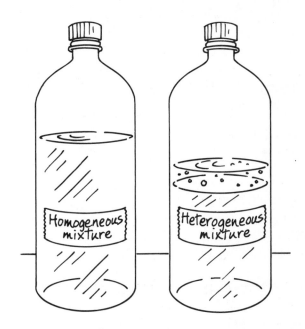

Results

The homogeneous mixture looks the same throughout, but the heterogeneous mixture has two separate layers.

Why?

When one or more substances are combined with another, they form a **mixture.** A mixture can be either **homogeneous** (the same throughout) or **heterogeneous** (not the same throughout). It is commonly said that oil and water do not mix, but any time oil and water are put in the same container and stirred or shaken, they combine to form a mixture. However, since the mixture separates into two visible layers after standing, it is heterogeneous. The salt **dissolves** (breaks up and thoroughly mixes with another substance) in the water, forming a **solution** (a homogeneous mixture in which one substance is dissolved in another) that doesn't separate after standing, but like all mixtures can be separated by other means, such as evaporation of the water leaving dry salt.

Separator

Benchmarks

By the end of grade 5, students should know that

• In something that is made of many parts, the parts usually influence each other.

By the end of grade 8, students should know that

• A system may be thought of as containing subsystems and as being a subsystem of a larger system.

In this investigation, students are expected to

• Separate a mixture into its parts.

• Identify changes that can occur in the physical properties of the ingredients of solutions, such as absorption of pigment in ink into filter paper.

Preparing for the Investigation

The best black ink to use is the ink from a water-soluble overhead projector pen.

Presenting the Investigation

1. Introduce the new science terms:

   absorb To soak up.

   attraction The ability to be drawn toward.

   chromatography A method of separating a mixture into its different substances.

   pigment A substance that gives color to material.

2. Explore the new science terms:

   • Pigment can be naturally found in living things or artificially made.

• The visible spectrum is made up of the rainbow colors of light. To remember these in the order in which they appear, use the mnemonic ROY G BIV: *R*ed, *O*range, *Y*ellow, *G*reen, *B*lue, *I*ndigo, *V*iolet.

• Pigments produce their color by absorbing certain colors of light and reflecting and/or transmitting others. The color that is *reflected* (bounced back) and/or *transmitted* (passed through) is the color you see.

• Absorbent paper, such as coffee filters, is used in chromatography.

• Chromatography depends on different factors, one of which is the attraction that the substances being separated have for absorptive paper.

• Black ink can be made by mixing red, blue, and yellow pigments with a liquid solvent.

Did You Know?

Ink print is a dehydrated *colloid*. A colloid is a mixture of a solid particles suspended in a fluid. Writing with ink leaves the wet colloid on the paper. When the fluid (liquid) evaporates, the solid particles are left on the paper.

EXTENSION

You may wish to have different groups use different brands of black pens and/or different-colored pens and compare the colors in the ink.

Separator

PURPOSE

To separate a mixture into its parts.

Materials

3 to 4 sheets of newspaper
3¼-by-2⅜-inch (8.1-by-5.9-cm) basket-type
  coffee filter
black water-soluble marker
10-ounce (300-ml) plastic cup
3-ounce (90-ml) paper cup
tap water

Procedure

1. Lay the newspaper on a table.

2. Spread the coffee filter on the newspaper.

3. Use the marker to draw a small
   flower in the center of the filter.

4. Set the plastic cup on the newspaper.

5. Stretch the filter over the
   mouth of the plastic cup.

6. Fill the small paper cup with
   water.

7. Dip your finger in the water, then touch the
   center of the flower with your wet finger.
   Observe the changes in the wet spot on the
   filter until the changes stop.

8. If there are any dry places in the flower drawn
   on the filter, repeat step 7.

Results

When water is added, the black ink begins to
separate into different colors. Depending on the
ink used, different amounts of red, blue, and yellow can be seen.

Why?

Ink is a mixture of a fast-drying liquid and
various coloring substances called **pigments.** The
pigment in the dried ink on the paper dissolves in
the water added to the paper. This watery mixture
is **absorbed** (soaked up) by and moves through
the paper. The different-colored pigments have
different amounts of **attraction** (ability to be
drawn toward) to the filter paper. The pigment
with the least attraction will move the greatest distance through the paper. Generally, blue pigment
in ink moves farthest, followed by yellow, then
red. This method of separating a mixture into its
different substances is called **chromatography.**

Forces and Motion

Matter is of course not *static* (at rest). It's always in motion, and there is no way to have motion without a force. In this section, forces and motion will be investigated.

Several kinds of forces are known to exist, including *gravitational* (the pull between objects), *friction* (a force caused by objects rubbing against each other), and *magnetic force* (the attraction between a magnet and a magnetic material). Forces can be balanced or unbalanced.

In this section, Newton's laws of motion will be used to describe and explain the differences between balanced and unbalanced forces. Students will also discover the turning effect of a force, called *torque,* and the balancing point of an object, called the *center of gravity* or *center of mass.*

Racers

Benchmarks

By the end of grade 5, students should know that

- A force causes an object to move, change speed, or stop moving. The greater the force, the greater the change in motion.

By the end of grade 8, students should know that

- An unbalanced force on an object changes the speed and/or direction of motion.

In this investigation, students are expected to

- Understand that changes in the speed of an object are caused by forces.
- Determine average speed.

Preparing for the Investigation

Students will need to work in groups of at least four. Students can use their watches to time each other. Calculators are optional. Students can do the simple arithmetic without a calculator.

Presenting the Investigation

1. Introduce the new science terms:

   average speed The total distance traveled divided by the total time.

   force A push or a pull on matter.

   motion The act or process of changing position.

   muscular force A force caused by changes in the length of muscles.

   speed The rate at which a distance is traveled in a given time.

2. Explore the new science terms:

   - A force is needed to cause motion.
   - A force can cause an object to move, change speed, or stop moving. The greater the force, the greater the change in motion. The more massive an object is, the more force it takes to move it.
   - When muscles shorten and lengthen, they provide muscular force that moves your body.
   - As force increases, motion increases.

- Speed is a measure of the time it takes to move a certain distance.
- Average speed is calculated over a period of time. For example, the speed is not the same throughout a trip. There are stops for traffic lights, lunch breaks, and so on. Average speed is calculated by dividing the total distance traveled by the total time it took to get there. For example, if you travel in a car 100 miles (160 km) in 2 hours, your average speed would be 100 (160) ÷ 2 = 50 miles (80 km) per hour.

Did You Know?

The cheetah can run about 70 miles (112 km) per hour for short distances of a few hundred yards (meters).

EXTENSION

Ask your students to prepare a bar graph to compare the average speeds of each person in their group. The numbering on the horizontal axis can represent average speed. For example, if the fastest speed is 22 inches (or 22 m) per second, then the horizontal numbering can be from 0 to 30. The names of each person in the group can be listed on the vertical axis, and bars filled in in different colors to show each person's average speed.

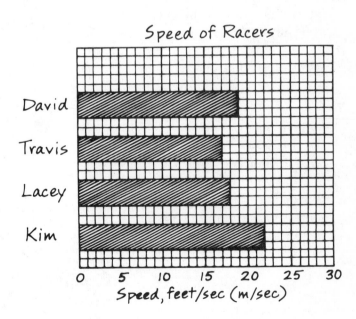

Speed of Racers

David

Travis

Lacey

Kim

0   5   10   15   20   25   30

Speed, feet/sec (m/sec)

Racers

PURPOSE

To determine the average speed of racers.

Materials

masking tape
pen
timer that counts seconds
yardstick (meterstick)
calculator (optional)

Procedure

1. Place a strip of masking tape on the floor to mark a starting line.

2. Write your name in row 1 of the Speed Data table.

3. Stand with your heels on the starting line.

4. Ask a helper to time you for 10 seconds while you walk as quickly as possible. When your helper says go, start walking. Stop and stand still as soon as your helper says stop.

5. Ask your helper to place a strip of tape on the floor at the heel of your forward foot.

6. Measure the distance between the two strips of tape in feet (meters).

7. Calculate your average speed by dividing your distance by the time, 10 seconds. Record your average speed in feet (meters) per second in the Speed Data table.

8. Ask each of the other helpers in your group to repeat steps 2 through 7.

Results

SPEED DATA	
Names of Racers	Average Speed
1.	
2.	
3.	
4.	

Why?

It takes **force** (a push or a pull) in order to move anything. The force used to move your body is called **muscular force** (a force caused by changes in the length of muscles). **Motion** (the act or process of changing position) is measured in **speed,** which is the rate at which a distance is traveled in a given time.

The total distance moved divided by the total time to travel this distance is called **average speed.** All of the racers in this activity used the same kind of motion, walking quickly, but some may have been able to move more quickly than others, so they had a greater average speed. One reason for differences in the average speeds of the racers could be the length of their legs. Those with longer legs would travel farther in 10 seconds.

Down!

Benchmarks

By the end of grade 5, students should know that

- Things near Earth's surface fall to the ground unless something holds them up.
- Earth's gravity pulls objects toward Earth's surface without touching the objects.

By the end of grade 8, students should know that

- Every object exerts gravitational force on every other object.

In this investigation, students are expected to

- Understand that gravity pulls things toward the center of Earth.

Preparing for the Investigation

The size and length of the dowel are not critical. Any straight rod, even a pencil or a ruler, will work.

Presenting the Investigation

1. Introduce the new science terms:

    gravity A force of attraction between all objects in the universe.

2. Explore the new science terms:

    - Some forces involve contact between objects. For example, lifting a fork involves contact between your hand and the fork.
    - Some forces act without one object being in contact with another object. Gravity, for example, can act over long distances, such as the gravity between the Sun and Earth, which keeps Earth revolving around the Sun.

- Falling objects or suspended (free-hanging) objects on or near Earth's surface are pulled toward the center of Earth by the force of gravity.
- Gravity pulls everything on or near Earth toward its center, so toward the center of Earth is what we really mean by "down."

Did You Know?

Astronauts' weightlessness isn't due to the fact that there is no gravity. They are weightless because their spacecraft is free-falling, meaning it is falling with gravity being the only force acting on it. You experience weightlessness each time you free-fall, such as when you dive off a diving board.

EXTENSION

On globes, the South Pole appears to be down, but really there is no down side of Earth. Have students help you prepare a bulletin board about gravity. You may wish to include a diagram showing children standing at different places on Earth playing with yo-yos that point "down" toward the center of Earth. You could include a picture of Galileo dropping objects from the Leaning Tower of Pisa, where he is said to have demonstrated experiments with gravity. You may also want to include a photograph of each student demonstrating an effect of gravity, such as standing on a bathroom scale, laying a book on a table, or walking. Next to the photograph, include a diagram made by the student, showing how things in the photo would look if there were no gravity.

Down!

PURPOSE

To understand the pull of gravity.

Materials

24-inch (60-cm) string
⅜-by-36-inch (0.94-by-90-cm) dowel
paper clip

Procedure

1. Tie one end of the string to the center of the dowel.
2. Secure the paper clip to the free end of the string.
3. Hold the dowel horizontally, one end in each hand, about 12 inches (30 cm) in front of your face. Observe the position in which the paper clip is hanging.

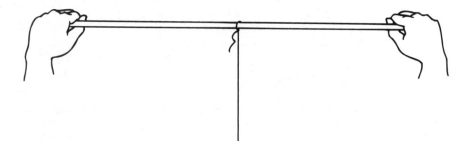

4. Tilt the dowel so that one end bends down and almost touches the string. Observe any change in the position in which the paper clip is hanging.

5. Return the dowel to a horizontal position and repeat step 4, tilting the other end down until it almost touches the string.

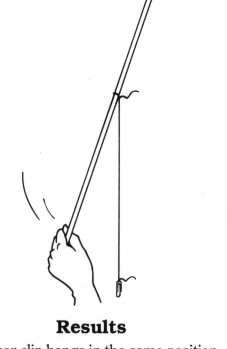

Results

The paper clip hangs in the same position regardless of which way the dowel is tilted.

Why?

The paper clip is free-hanging, meaning that it is held in place only by the string and is free to swing in any direction. Earth's **gravity** (a force of attraction between all objects in the universe) pulls the paper clip down, toward the center of Earth. When you tilt the dowel, the string does not tilt with the rod, so the paper clip continues to be pulled in the same direction by gravity. On or near Earth's surface, gravity pulls everything toward the center of Earth.

More or Less?

Benchmarks

By the end of grade 5, students should know that

- Earth's gravity, like gravity on other planets, pulls on objects.

By the end of grade 8, students should know that

- Other planets have compositions and conditions very different from Earth's, including their gravitational force.

In this investigation, students are expected to

- Compare the gravity of different planets.
- Identify the effect of gravity on weight.

Presenting the Investigation

1. Introduce the new science terms:

   celestial bodies Natural objects in the sky, such as planets, moons, stars, and suns.

   gravity rate (G.R.) The surface gravity of a celestial body divided by the surface gravity of Earth, thus Earth's G.R. is 1.

   newton (N) The SI unit of weight.

   pound The English unit of weight.

   surface gravity Gravity at or near the surface of a celestial body.

   weight A measure of the force of gravity, which on Earth is a measure of the force with which Earth's surface gravity pulls on an object.

2. Explore the new science terms:

   - Gravity is the strength with which an object is pulled toward another object. The strength of Earth's surface gravity in SI units is 9.8 N/kg. This means that Earth's gravity pulls on objects with a force of 9.8 N (newtons) for every kilogram (kg) of mass in the object.

- What we call *weight* is a measure of the force with which Earth's gravity pulls on an object, such as a person's body.

- Gravity rating of Earth is usually referred to as 1g and is read as it is written—1 g—and is understood to mean $1 \times$ Earth's surface gravity (g) of 9.8 N/kg. So, 1.16 g for Saturn means its surface gravity is 1.16×9.8 N/kg. The 1 g for Earth is not to be confused with the SI measurement of mass of 1 g, which is read as 1 gram. The difference can be determined by the content of the material being read.

- Weight, also called force weight (F_{wt}), is the product of an object's mass times the strength of gravity. This is expressed by the formula $F_{wt} = m \times g$.

- On Earth, 1 pound equals 4.5 N.

- The force of gravity depends on the mass of two objects and the distance between their centers. Gravity increases with mass and decreases with distance between the centers of two objects.

- An object's weight varies on different planets because planets have different surface gravity.

EXTENSION

The mass of an object would not change even though its weight would vary on another planet. Have students compare their mass on one or more planets with their mass on Earth. Their mass can be determined by dividing their weight on the planet in question by the gravitational strength of the planet. This is expressed as $m = F_{wt} \div g_{planet}$.

Note: The gravitational strength of the planets can be determined by multiplying Earth's gravitation strength (9.8 N/kg) by the G.R. of the planet in question. This is expressed as $g_{planet} = g_{Earth} \times G.R._{planet}$.

More or Less?

PURPOSE

To determine what your weight would be on different planets.

Materials

bathroom scale
pencil
calculator

Procedure

1. Determine your weight in pounds by weighing yourself on the scale.

2. Calculate your weight in newtons (N) by using the calculator to multiply your weight in pounds by 4.5. For example, if you weigh 90 pounds, your weight in newtons would be:

90 pounds × 4.5 = 405 N

3. Determine what your weight would be on the different planets by using the calculator to multiply your weight on Earth by each planet's "gravity rate" listed in the table. For example, if you weigh 405 N on Earth, your weight on Mars, which has a gravity multiple of 0.38, would be:

405 N × 0.38 = 153.9 N

Results

You weight will vary on each planet. You would weigh the least on Pluto and the most on Jupiter.

Why?

Your **weight** is a measure of the force of gravity, which on Earth is a measure of the force with which Earth's surface gravity pulls on an object. **Surface gravity** is gravity at or near the surface of a **celestial body** (a natural object in the sky, such as a planet, moon, star, or sun). The **gravity rate (G.R.)** of a celestial body is its surface gravity divided by Earth's. Earth's G.R. equals 1. Celestial bodies with a gravity rate greater than 1 have a surface gravity greater than that of Earth. Those with a gravity multiple less than 1 have a surface gravity less than that of Earth. As shown in this investigation, your weight on each planet is calculated as the product of your weight on Earth times the gravity rate of each planet. The **pound** is the English unit of weight; the **newton (N)** is the SI unit of weight. Weight on celestial bodies in this investigation is measured in newtons.

WEIGHT DATA		
Planet	**Gravity Rate (G.R.)**	**Weight on Planet in newtons (G.R. × weight on Earth)**
Mercury	0.38	
Venus	0.90	
Mars	0.38	
Jupiter	2.54	
Saturn	1.16	
Uranus	0.92	
Neptune	1.19	
Pluto	0.06	

Force Pattern

 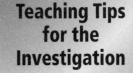 

Benchmarks

By the end of grade 5, students should know that

- Magnets can be used to move objects made of iron without touching the objects.

By the end of grade 8, students should know that

- A force, such as magnetism, can move some objects.

In this investigation, students are expected to

- Identify the magnetic lines of force around a magnet.
- Understand that magnetism is a physical property of some substances.

Preparing for the Investigation

If you are doing the investigation with young children, you should cut the pipe cleaners in advance.

Presenting the Investigation

1. Introduce the new science terms:

   magnet An object that is surrounded by a magnetic field and attracts magnetic materials.

   magnetic field The space around a magnet where a magnetic force can be detected.

   magnetic force The attraction between magnets or between a magnet and a magnetic material.

   magnetic lines of force A pattern of lines representing the magnetic field around a magnet.

   magnetic materials Materials that can be attracted to or magnetized by a magnet, such as iron and steel.

   magnetic pole One of two ends of a magnet where the magnetic field is strongest.

   magnetism Magnetic force.

2. Explore the new science terms:
   - Magnetic materials include iron, cobalt, nickel, and mixtures of these metals with each other and/or with other substances. (Steel is a mixture of iron, carbon, and various metals.)
   - The magnetic field around a magnet is made of lines of force that move out of the north pole, around the magnet, and into the south pole of the magnet.
   - The magnetic field around Earth is called the *magnetosphere*.
   - The magnetic ends of Earth are called the *magnetic north and south poles* of Earth.
   - When a bar magnet is suspended, the pole that points toward Earth's magnetic north pole is the north-seeking or *north pole of the magnet*. The opposite end of the bar magnet points toward Earth's magnetic south pole and is called the south-seeking or *south pole of the magnet*.
   - On a disk magnet, the flat surfaces are the poles.
   - The attraction and repulsion between two magnets decreases as the distance between their poles increases.

Did You Know?

Because Earth's geographic and magnetic poles are in different places, a compass doesn't point to *true north* (defined as the north end of Earth's axis, an imaginary line around which Earth rotates). Instead, it points due north, toward Earth's north magnetic pole.

EXTENSION

Another way to show the magnetic lines of force of a magnet is to put 1 teaspoon (5 ml) of iron filings inside a resealable plastic bag. Iron filings are found at teacher supply stores or from science suppliers in appendix 3. Hold the bag so that the iron filings are just above the magnet. Gently shake the bag to spread the filings. The iron filings will line up with the magnetic field around the magnet. Repeat the investigation with magnets of different shapes (bar, horseshoe, etc.).

Force Pattern

PURPOSE

To produce a pattern representing the magnetic field around a disk magnet.

Materials

scissors
12-inch (30-cm) pipe cleaner
disk magnet
index card
school glue

Procedure

1. Cut the pipe cleaner in half. Then cut each piece in half. Continue cutting the pieces in half until you have 16 equal-size pieces.

2. Place the magnet on a table and cover it with the index card.

3. Spread a thick layer of glue over the top of the card in the area that covers the magnet.

4. Hold the pipe cleaner pieces about ½ inch (1.25 cm) above the glue on the card, with the ends pointing straight down, then drop them one at a time into the glue.

5. Keep the pipe cleaner pieces and card in place until the glue dries, which should take about 1 hour.

Results

The pipe cleaner pieces make a round design in which the outer pieces lean outward at an angle and those in the center stand more vertically. The glue keeps the pipe cleaner pieces in place.

Why?

Magnetic force, or **magnetism,** is the attraction between magnets or between a magnet and a magnetic material. A **magnet** is an object that is surrounded by a **magnetic field,** the space where magnetic force can be detected. **Magnetic materials** are those which can be attracted to or magnetized by a magnet, such as iron and steel. The magnetic field extends from one end, or **magnetic pole,** of the magnet to the other, and is strongest at the poles. The poles of a disk magnet are its flat surfaces. Pipe cleaners contain a thin steel wire, which is attracted to the magnet. When the steel in the pipe cleaners enters the magnetic field, the steel is pulled toward the magnet. The arrangement of the pipe cleaner pieces indicates the direction of the invisible **magnetic lines of force** (the pattern of lines representing the magnetic field around a magnet) around the upturned pole of the disk magnet.

Janice VanCleave's Teaching the Fun of Science

Tug-of-War

Benchmarks

By the end of grade 5, students should know that

- Without touching either magnetic materials or other magnets, a magnet pulls on magnetic materials and either pushes or pulls on other magnets.

By the end of grade 8, students should know that

- If the forces are balanced, an object may remain motionless even though forces are acting on it.

In this investigation, students are expected to

- Understand that forces, including magnetic forces, can be balanced or unbalanced.
- Demonstrate how an unbalanced force causes an object to move.

Preparing for the Investigation

The magnets used by each student or group need to look alike. Disk magnets about the size of a quarter are inexpensive and easy to slide along the ruler, but any shape will work. Point out that although it is assumed that the magnets used are of equal strength, the results of the investigation will determine whether this is true.

Presenting the Investigation

1. Introduce the new science terms:

   balanced force Force applied equally to an object from opposite directions.

   unbalanced force Force applied on an object without an opposing force of equal strength.

2. Explore the new science terms:
   - When a balanced force is applied to an object at rest, the object remains at rest.
   - When a balanced force is applied to an object in motion, the object continues to move at the same speed and in the same direction.
   - An unbalanced force causes an object at rest to start moving.
   - An unbalanced force causes a moving object to change speed and/or direction of motion. In other words, the object *accelerates* (increases in speed or changes direction) or *decelerates* (decreases in speed) depending on whether the opposing force is increased or decreased, respectively.
   - Forces acting in the same direction on an object along a straight line add to each other.
   - Forces acting in opposite directions on an object along a straight line cancel each other.
   - Like poles of a magnet repel each other, and unlike poles attract.
   - The strength of a magnetic field decreases with distance of the magnet from an object.

Did You Know?

Maglev, or magnetic levitation, trains work on the principle that like poles of a magnet repel each other. Magnets on the track and other magnets on the train allow Japan's HSST (High Speed Surface Transport) maglev train to *levitate* (rise) about 4 inches (10 cm) above the track as it rockets along at up to 300 miles (480 km) per hour.

EXTENSION

Repeat the experiment, using magnets of different shapes.

Tug-of-War

PURPOSE

To use balanced and unbalanced magnetic forces to compare the strengths of two magnets.

Materials

4-foot (1.2-m) string
paper clip
transparent tape
yardstick (meterstick)
2 identical disk magnets

Procedure

1. Tie one end of the string to the paper clip.
2. Use a piece of tape to secure the free end of the string to the edge of a table so that the paper clip hangs about 1 inch (2.5 cm) above the floor.
3. Lay the measuring stick on the floor under the hanging paper clip so that the paper clip is just above the middle of the measuring stick.
4. Place a magnet on each end of the measuring stick. The magnet at the zero end of the

measuring stick is magnet A and the one at the opposite end is magnet B.

5. Slowly slide magnet A along the measuring stick toward the paper clip. Continue until the paper clip moves toward the magnet. Stop and make note of the distance the magnet is from the center of the measuring stick in the Magnet Distance Test Data table.
6. Repeat step 5 three more times and average the test distances for the magnet.
7. Repeat steps 5 and 6 using magnet B.
8. Based on the averages, decide whether the magnets are of equal strength or one magnet is stronger.
9. Test your conclusion in step 8 by placing each magnet at the calculated average distance.

MAGNET DISTANCE TEST DATA					
Magnet	Test 1	Test 2	Test 3	Test 4	Average
A					
B					

Results

The results will vary depending on the strengths of the magnets used.

Why?

If the magnets are of equal strength, the magnetic forces will be balanced and the distances from the paper clip will be equal. A **balanced force** is one that is applied equally to an object from opposite directions. If the magnets are not of equal strength, their combined force will be unbalanced and the paper clip will move toward the stronger magnet when both magnets are an equal distance away. An **unbalanced force** does not have an opposing force of equal strength.

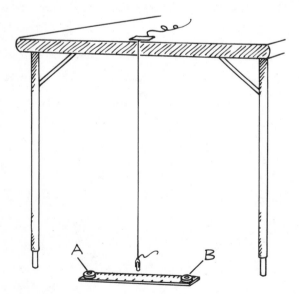

Clinger

Benchmarks

By the end of grade 5, students should know that

- Material that has been electrically charged pulls on all other materials and may either push or pull other charged materials without touching them.

By the end of grade 8, students should know that

- An unbalanced force causes a moving object to change speed and/or direction of motion.

In this investigation, students are expected to

- Identify the effects of static electricity.

Preparing for the Investigation

If puffed rice cereal is not available, use small pieces of paper or Styrofoam.

Presenting the Investigation

1. Introduce the new science terms:

   electron A negatively charged particle outside the nucleus of an atom.

   nucleus The center of an atom.

   proton A positively charged particle in the nucleus of an atom.

   static electricity A buildup of electric charges, either positive or negative.

2. Explore the new science terms:

   - All matter is made up of atoms. Atoms have a center, called a nucleus, which contains positively charged particles called protons. Spinning outside the positively charged nucleus are negatively charged particles called electrons. Generally, when any two materials are rubbed together, electrons tend to be rubbed off one of the materials and onto the other. This causes one of the materials to be more positively charged and the other more negatively charged. This buildup of charges is called static electricity.

   - When your feet rub against a carpet, electrons from the carpet are rubbed off onto your feet. Opposite charges attract. So as you get close to another object, especially a doorknob or other metal object, the protons in the doorknob attract the extra electrons in your body. The electrons travel through your body and move from your hand to the doorknob when you reach for it, causing a small electric shock.

Did You Know?

There is more static electricity in autumn and winter than in spring and summer. This is because there is more static electricity when the air is cold and dry, as it is in autumn and winter. Warm air can hold more moisture. When the air is moist, the water molecules in the air pick up electrons, preventing them from collecting in your body.

EXTENSION

Use the attraction between opposite charges to create a moving butterfly model. Have each student draw a butterfly on a 4-inch (10-cm) -square piece of tissue paper. Cut out the butterfly design. Put a small amount of glue on the underside of the butterfly's body, and glue the butterfly to a 6-inch (15-cm) -square piece of cardboard. Make sure the wings are not glued down. Allow the glue to dry. Crease the wings alongside the body so they bend up and down easily. Charge the food wrap as before or charge an inflated balloon (make sure the balloon is dry) by rubbing it on your hair. Hold the charged balloon near but not touching the wings, then move it away. Repeat this motion of the balloon so that the butterfly wings flutter up and down.

Clinger

PURPOSE

To demonstrate static electricity.

Materials

20 to 25 pieces of puffed rice cereal
2-foot (60-cm) piece of plastic food wrap
sheet of typing paper

Procedure

1. Put the pieces of cereal on a table.
2. Crumple the plastic wrap into a fist-size wad.
3. Quickly rub the wad of plastic wrap back and forth across the sheet of paper 10 to 15 times. Immediately hold the plastic wrap above the cereal pieces, near, but not touching, the cereal.

Results

The cereal leaps up to the plastic.

Why?

Atoms have a center called a **nucleus,** which contains positively charged particles called **protons.** Spinning outside the nucleus are negatively charged particles called **electrons.** Generally, when any two materials are rubbed together, such as the plastic and the paper, electrons are lost from one material (the paper) and gained by the other (the plastic). The buildup of electric charges on an object is called **static electricity** because the charges are stationary (nonmoving). When the negatively charged plastic approaches the cereal, the positive charges in the cereal are attracted to the negative charges in the plastic. This attraction is great enough for the lightweight cereal to break free of the downward pull of gravity and move upward, sticking to the plastic.

Chop!

Benchmarks

By the end of grade 5, students should know that

- The greater the force, the greater the change in motion of the object to which the force is applied.
- The more massive the object, the less the effect of a given force on it.

By the end of grade 8, students should know that

- Objects have a tendency to resist a change of motion. Those at rest remain at rest and those in motion remain in motion unless acted on by an outside force.

In this investigation, students are expected to

- Demonstrate that an object will remain at rest due to its inertia.
- Identify the effect of mass on the state of inertia of an object.

Preparing for the Investigation

If sand is not available, use any material that will add weight to the cups, such as aquarium gravel.

Presenting the Investigation

1. Introduce the new science terms:

   friction A force that opposes the motion of one object whose surface is in contact with another object.

   inertia The tendency of an object to remain at rest or to resist any change in its state of motion unless acted on by an outside force.

2. Explore the new science terms:
   - An object's state of motion is caused or changed by outside forces.

- Because of inertia, an object will not change its motion unless an unbalanced force acts on it.
- Because of inertia, an object at rest stays at rest.
- Because of inertia, a moving object continues to move in a straight line at a constant speed. For example, put a marble on a flat surface and give it a push. This unbalanced force causes the marble to move. Because the marble has inertia, it continues to move after you stop pushing it. But another unbalanced force, friction, acts on the marble, causing it to slow and finally stop.
- The more massive the object, the greater its inertia. The force needed to move an object increases with the object's inertia.
- Friction acts in a direction opposite to the motion of the object.
- The rougher the two surfaces rubbing against each other, the greater the friction.

Did You Know?

Sand is spread on icy roads and walkways to increase friction. You are less likely to fall on sanded ice than on unsanded ice.

EXTENSION

Ask the class to think about how an entertainer can pull a tablecloth off a table and leave the items in place on the cloth. (The entertainer must use heavy items on the table to increase the inertia of the items and a slick tablecloth to reduce friction. A large unbalanced force must also be applied to the tablecloth.)

Chop!

PURPOSE

To determine how mass affects the inertia of an object.

Materials

two 10-ounce (300-ml) transparent plastic cups
sand
ruler
2-by-12-inch (5-by-30-cm) strip of waxed paper

Procedure

1. Fill one of the cups with sand.
2. Lay about 4 inches (10 cm) of the strip of waxed paper near the edge of a table.
3. Set the empty cup on the end of the waxed paper strip.
4. Try to pull the strip out from under the cup without moving the cup. Do this by holding the free end of the strip and hitting the paper just past the table edge with the edge of your hand like a karate chop as shown. Observe any motion in the cup.
5. Repeat steps 2 to 4 three or more times.
6. Repeat steps 2 to 4 four times, using the cup of sand.

Results

The paper moved from under both cups, but the cup of sand remained in place or moved less than the empty cup.

Why?

Waxed paper is slippery, so there is little **friction** (a force that opposes the motion of one object whose surface is in contact with another object) between the cups and the paper. The cup of sand has more mass than the empty cup. Since the more massive cup remained in place or moved less, it has more **inertia** (the tendency of an object to remain at rest or to resist any change in its state of motion unless acted on by an outside force). The more massive the object, the greater its inertia.

Energy

Matter can be seen and touched, but *energy* cannot. You can see the effects of energy by observing matter. For example, energy makes things move. When you see a football flying through the air, you know that energy was used to throw it.

Energy and matter are interchangeable. Under familiar conditions, matter never changes, but in nuclear changes, as in the Sun, matter is changed to energy. The laws of conservation of *matter* and *energy* state that matter and energy cannot be created or destroyed. They may change from one phase or form to another, but the total amount of energy and matter in the universe remains constant.

In their investigations of energy, students will discover that energy exists in different forms, including *heat, light, sound,* and *electricity.* These forms are divided into two basic groups—*potential* and *kinetic.*

Benchmarks

By the end of grade 5, students should know that

- Energy is necessary for motion.

By the end of grade 8, students should know that

- Energy cannot be created or destroyed, but only changed from one form to another.
- Energy appears in different forms.

In this investigation, students are expected to

- Identify the effect that height has on the energy of an object.
- Determine the differences between the two basic groups of energy, potential and kinetic.
- Determine that energy is necessary to do work.

Preparing for the Investigation

Socks must be clean white socks with no holes. Use any scale or balance that measures in gram units.

Presenting the Investigation

1. Introduce the new science terms:

   energy (E) The ability to do work.

   gravitational potential energy (GPE) Potential energy due to an object's height above a surface.

   joule (J) An SI unit by which work is measured; $1N \cdot m = 1J$.

   kinetic energy (KE) Energy that a moving object possesses because of its motion.

   potential energy (PE) Stored energy of an object due to its position or condition.

   work (w) The movement of an object by a force.

2. Explore the new science terms:

   - Work is the product of a force on an object times the distance the force moves the object. For example, a box weighing 1N is moved 1m. Work is determined by this equation: $w = f_{wt} \times d = 1N \times 1m = 1N \cdot m = 1$ joule. The dot between the units indicates that they have been multiplied together. (It is also correct to express the units as Nm—without the dot.)
   - When work is done on an object, the object gains energy equal to the amount of work done on it. For example, if work of $1N \cdot m$ is done on an object, it gains 1 joule of energy.

- If an object does work, it loses energy equal to the work is has done.
- All forms of energy can be divided into two groups, potential energy and kinetic energy.
- An object at rest that is capable of movement because of its position or condition possesses potential energy.
- Examples of objects having potential energy of position are water behind a dam or a rock on the edge of a high cliff. Examples of objects having potential energy of condition are a stretched rubber band or a compressed spring. None of these objects is moving, but all have the potential of moving and doing work.
- If an object's position is above the ground, the object is said to have gravitational potential energy.
- As an object falls, its gravitational potential energy changes to kinetic energy.
- When potential energy changes to kinetic energy, matter moves.
- Just as potential energy can be changed to kinetic energy, kinetic energy can be changed to potential energy.
- The work done to raise an object equals the gravitational potential energy it has at rest at the higher position. This potential energy is equal both to the kinetic energy the object has when it hits the surface it was raised above and to the work done on the surface.
- Examples of objects having kinetic energy are a moving car or a waterfall.

Did You Know?

One pint (0.5 L) of water with a mass of 454 g has about 4,271 J of GPE at the top of Angel Falls in Venezuela, which is the tallest waterfall in the world, at a height of over 3,200 feet (960 m).

EXTENSIONS

Repeat the experiment, using socks with different amounts of rice in them. Use a marker to number the socks so that those with the same amount of rice have the same number.

Equal

PURPOSE

To determine the effect that height has on the gravitational potential energy of an object.

Materials

1 cup (250 ml) dry rice
sock
food scale that measures grams
calculator
pen
yardstick (meterstick)

Procedure

1. Pour the rice in the sock and tie the sock.
2. With the food scale, measure the mass of the sock of rice to the nearest gram (g).
3. Determine the weight of the sock of rice in newtons (N) by using the calculator and the following equation. Record the weight in the "Force Weight" column of the Energy Data table.

$$\text{Force weight} = \text{mass} \times 0.0098 \text{ N/g}$$

   The force needed to lift the sock is equal to its weight, which can be called the force weight (f_{wt}).

4. Calculate the work (w) done if the sock were lifted a height or distance (d) of 0.5 m, using the following equation. Record the answer in the "Work" column of the Energy Data table.

$$w = f_{wt} \times d$$

   Note: The force unit of newton (N) times the distance unit of meter (m) equals N • m, which is equal to the work unit of joule (J).

5. Ask a helper to raise the sock to a height of 0.5 m above the floor.
6. Holding your hand palm up just above the floor and in line with the sock, ask your helper to release the sock. Note how it feels when the sock hits your hand. Record your observations in the table.

ENERGY DATA			
Force Weight (f_{wt}), n	Height (d), m	Work (w), J	Observations
	0.5		
	1.0		

7. Repeat steps 4 to 6, using a height of 1 m.

Results

The higher the sock is held, the greater the work done in lifting it and the greater the energy the sock has when dropped. The sock hits your hand harder when it is dropped from a higher height.

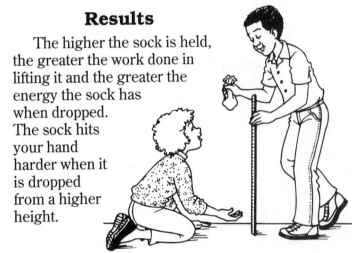

Why?

Energy (E) is the ability to do **work (w)**, which occurs when a force moves an object. **Potential energy (PE)** is stored energy. When an object is raised above a surface, it is said to have **gravitational potential energy (GPE).** The higher the object is raised, the greater its GPE. GPE is also equal to the work done to raise the object, and equal to the work the object can do when it drops from its raised position. As the object falls, its GPE changes to **kinetic energy (KE)** (the energy of a moving object). In this investigation, the higher the sock was raised, the greater the work that was done to lift it, the greater the gravitational potential energy it had at its height, and the greater the kinetic energy it had when it hit your hand. Work and energy in this investigation are measured in **joules (J).** One joule is the amount of work done when a force equal to 1 N is applied over a distance of 1 m.

No Loss

Benchmarks

By the end of the grade 5, students should know that

- Energy can be transferred from one object to another.

By the end of grade 8, students should know that

- Energy cannot be created or destroyed, but only changed from one form to another.
- Energy appears in different forms. Mechanical energy exists in moving bodies and bodies capable of motion.

In this investigation, students are expected to

- Demonstrate and identify energy transfer.

Preparing for the Investigation

Students will need to work in pairs. You may wish to punch holes in the lids in advance. Use one lid for each group of two students. The plastic lids can come from food cans.

Presenting the Investigation

1. Introduce the new science terms:

   law of conservation of energy A law of physics that states that energy can be changed from one kind to another, but cannot be created or destroyed under normal conditions.

   law of conservation of mechanical energy A law of physics that states that the sum of the potential and kinetic energy of an object remains the same as long as no outside force acts on it.

   mechanical energy Energy of motion; sum of the potential and kinetic energy of an object.

2. Explore the new science terms:

   - Water running downhill in a stream has *kinetic mechanical energy*. Water in a water tower high above the ground is not moving, but is capable of motion. Because of its position it has gravitational potential energy, which can also be called *potential mechanical energy*.

   - An object with mechanical energy does not have to be in motion but must have potential energy, which has the ability to produce motion.

   - Energy can be changed from one form to another. For example, the electrical energy of lightning changes into light energy that can be seen, heat energy that can be felt, and sound energy that can be heard.

   - The total energy in the universe is constant. All the different forms of energy in the universe add up to the same total amount of energy at all times. But because energy changes from one form to another, there may be more or less of one form of energy at any given time.

   - The law of conservation of mechanical energy states that disregarding any outside force, such as friction, the sum of the potential and kinetic energy of a substance remains the same. For example, if the total mechanical energy equals 40 J, the potential energy at the top of the tower is 40 J and the kinetic energy is 0 J. Halfway between the top and the bottom of the tower, the potential energy equals 20 J and the kinetic energy equals 20 J. When the water hits the ground at the bottom of the tower, its kinetic energy equals 40 J and its potential energy equals 0 J.

Did You Know?

Generally, no matter how efficiently one form of energy is converted to another, some energy is always lost as heat in the process.

EXTENSION

Have students replenish the energy of the system. Repeat step 5 of the procedure, but when the lid starts to rewind the string, do not hold the string taut. This allows the string to wind more. Pull the string taut again until the lid starts to rewind the string, then loosen the string again, then tighten, and so on. The lid will continue to spin back and forth.

Up and Down

PURPOSE

To read a thermometer.

Materials

model of a thermometer (provided by teacher, or see instructions or separate sheet)

Procedure

1. Look at the scale on the model thermometer and determine the value of each division.

2. Move the colored strip so that the top of the strip is in line with the sixth division up from zero on the scale. Determine the temperature at this mark.

3. Move the colored strip so that it is between the ninth and tenth divisions on the scale. Determine the temperature between these marks.

Results

Temperature readings of 6°C and 9.5°C are found on the model thermometer.

Why?

A **thermometer** is an instrument used to measure **temperature** (the physical property that determines which direction heat flows between substances). Basically a thermometer detects how fast the particles of a material are moving. The faster the motion of the particles, the greater the temperature reading.

The model thermometer in this experiment represents a portion of a Celsius thermometer. On the **Celsius scale,** the freezing point of water is 0° and the boiling point is 100°. The model represents a thermometer made of a long tube with liquid inside. When the liquid is warmed, it expands and moves up the tube. The reading on the scale gets higher as the temperature increases. When the liquid is cooled, it contracts and moves down the tube. As the temperature decreases, the reading on the scale decreases. For the scale on the model, each division has a value of 1 degree Celsius (1°C). So a point halfway between two divisions has a value of half a degree, which can be written 0.5°C.

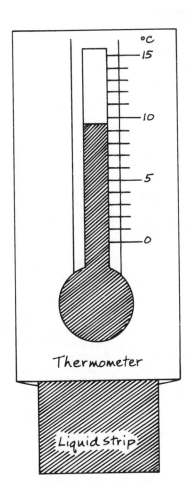

Thermometer

Liquid strip

Benchmarks

By the end of grade 5, students should know that

- Things get bigger when they are heated.

By the end of grade 8, students should know that

- In a fluid material, currents can develop that aid the transfer of heat.

In this investigation, students are expected to

- Describe methods of heat transfer in fluids.
- Observe convection currents in a fluid.

Preparing for the Investigation

Collect and clean empty soda bottles before the experiment. You may want to cut the tops off the bottles and cover the rough cut edges of the bottles with masking tape yourself, especially for young children. Even for older children, make a cut in each bottle with a knife so they have a place to safely start cutting.

Presenting the Investigation

1. Introduce the new science terms:

    convection The transfer of heat from one region to another by the circulation of currents in a fluid.

    convection currents The circular movement of fluids of unequal temperature.

2. Explore the new science terms:

    - Heat can be transferred by conduction (the collision of particles) through any phase of matter. The particles, with heat added, may change in degree of vibration, but stay in relatively the same place. For example, when a metal rod is heated, the parti-

    cles in the rod vibrate faster, but the rod does not move across the room.

    - Heat can be transferred through fluids (gases and liquids) by convection. In this type of heat transfer, the material does move from one place to the other. Heated fluids rise and cooled fluids sink, creating convection currents.

Did You Know?

- Most ocean currents flow in one direction all the time, but in the northern Indian Ocean, surface currents change direction twice a year, driven by the monsoon winds. Part of the year the currents move away from India, and the other part of the year the currents move toward India.
- While sailing south near the northeastern coast of Florida, Spanish explorer Juan Ponce de Léon (1460–1521) discovered that he was making no progress because the water was flowing in a northerly direction. He was sailing in a current that was later to be called the Gulf Stream. In 1769, American statesman and scientist Benjamin Franklin (1706–1790) published a chart of the Gulf Stream.

EXTENSION

Ask your students how using cold water in the small jar and warm water in the plastic bottle might affect the results. Demonstrate either by repeating the experiment, reversing the temperature of the water in the containers, or by allowing the students to devise an experiment themselves.

Movers

PURPOSE

To observe convection currents in water.

Materials

scissors
clean, empty, 2-liter plastic soda bottle
masking tape
2 cups (500 ml) crushed ice
tap water
small jar
spoon
blue food coloring
6-inch (15-cm) piece of aluminum foil
rubber band
pencil
timer

Procedure

1. Cut the top third off the 2-liter bottle and cover the rough edges of the bottom section with masking tape. Discard the top section.

2. Place the ice in the bottom section of the bottle. Fill the bottle about half full with cold water.

3. Fill the small jar to the brim with warm water. Use the spoon to stir in 6 to 8 drops of food coloring.

4. Cover the mouth of the small jar with the aluminum foil. Use the rubber band to secure the foil around the mouth of the jar.

5. Lower the small jar into the ice water in the bottom section of the bottle.

6. Remove any unmelted ice from the water. The plastic bottle should be at least three-fourths or more full of water. Add cold water if necessary.

7. Use the point of the pencil to make two holes in the aluminum foil covering the small jar.

8. Observe the surface of the foil, the water above the foil, and the contents of the small jar for 3 minutes or until no further changes are seen.

Results

The colored water rises from one of the holes in the jar's foil covering, but the jar remains full. The color of the water in the jar looks lighter as time passes.

Why?

Water molecules, like the molecules of all fluids, are spaced closer together when cool and farther apart when warm. Warm water therefore weighs less than an equal volume of cool water. The warm colored water rises through one of the holes in the foil covering the jar, and the cool clear water sinks and enters the jar through the other hole to take the place of the warm water that left. This circular movement of fluids of unequal temperature, called a **convection current,** continues until the water inside the jar is the same temperature as the water outside. The transfer of heat from one region to another by the circulation of currents in a fluid is called **convection.**

Radiate

Benchmarks

By the end of grade 5, students should know that

- Things that give off light often also give off heat.
- A warm object can warm a cool object from a distance.

By the end of grade 8, students should know that

- Heat can be transferred by radiation.

In this investigation, students are expected to

- Describe a method of heat transfer in space.
- Verify that radiant energy doesn't need matter to be transferred.

Preparing for the Investigation

On a sunny day, you may wish to take the class outdoors and use the heat of the Sun instead of a desk lamp. If so, follow the same procedure, but have students position their hands so that the Sun shines directly on the top hand. *Caution students not to look at the Sun because it could damage their eyes.*

Presenting the Investigation

1. Introduce the new science terms:

   electromagnetic wave A disturbance in electric and magnetic fields; disturbance that can move through space.

   infrared radiation Radiation that all objects give off and that produces heat when absorbed.

   radiant energy A form of energy that travels in electromagnetic waves.

   radiation Radiant energy; also the transmission of radiant energy in waves.

2. Explore the new science terms:
   - *Waves* are disturbances that move through matter or space.
   - Waves are created by motion, such as ripples (waves) across a pond produced by dropping a stone in the water. While most waves are a disturbance of matter, electromagnetic waves are produced by the motion of electrons and are a disturbance of magnetic and electric force fields.

- Radiant energy does not require matter to move from one place to another, so radiant energy can travel from the Sun through space to Earth.
- Some forms of radiation can be felt as heat, such as infrared; others can be seen, such as visible light; and some can pass through your body without being felt or seen, such as *X rays*.
- All forms of radiant energy travel at the speed of light, 186,000 miles (300 million m) per second.
- All objects *absorb* (take in) and *radiate* (give off) infrared radiation. The warmer the object, the more infrared radiation it gives off. Thus, infrared radiation is often called *heat energy*. Even ice radiates infrared radiation, although in a very small amount.
- Radiation does not physically transfer heat as do conduction and convection. Instead radiation, when absorbed by an object, causes the object to get hotter.
- Radiation from the Sun is called *solar radiation* and is made of radiation of different-size electromagnetic waves. Most solar radiation is visible light, ultraviolet radiation, and infrared radiation.

Did You Know?

If the nuclear reactions in the Sun's core stopped today, it would be about 10 million years before the Sun's surface cooled enough for Earth to be affected.

EXTENSION

Microwaves are also a type of radiant energy. In a microwave oven, only *polar molecules* (those with positive and negative ends), such as molecules of water, get hot. In a microwave, polar molecules are flipped back and forth. In the process, the molecules bump into each other, and the friction of this motion produces heat energy. To demonstrate how friction causes materials to heat up, ask students to rub their hands together very quickly. The faster they rub, the warmer their hands will feel.

Radiate

PURPOSE

To demonstrate that heat can travel by radiation.

Materials

ruler
desk lamp

Procedure

1. Hold one hand about 2 inches (5 cm) under the other.
2. Move your hands so that the top hand is 6 inches (15 cm) below the bulb of the desk lamp.
3. Hold your hands in this position for about 5 seconds. Make note of how warm or cool each hand feels.

Results

The top hand, which is closer to the light, feels much warmer than the bottom hand.

Why?

Radiant energy is energy that travels in **electromagnetic waves** (a disturbance in electric and magnetic fields). All objects give off **infrared radiation,** which produces heat when absorbed. Radiant energy as well as its transmission is called **radiation.**

In this investigation, your hand is heated basically by infrared radiation given off by the hot lightbulb, not by the conduction of heat from the bulb through the air. Air molecules around the bulb are warmed by conduction, but most of the heated air spreads out and rises toward the ceiling. Since only the top hand feels warmer, the extra heat felt is not due to air touching the skin but to the infrared radiation. The top hand absorbs the infrared radiation, and the lower hand does not receive it. When your hand absorbs this radiation, the molecules in your skin move faster, thus your skin has more thermal energy and feels warmer.

If the air around it were warmer, your hand would feel warmer faster. The extra heat felt by the top hand is not due to air touching it. Heat in the form of infrared radiation from the lightbulb is transferred to your hand by radiation. When your hand absorbs this radiant energy, it feels warmer.

Life Science

Life science is the study of the way living organisms behave and interact. Life science focuses on three major areas: *botany*, the study of plants; *zoology*, the study of animals; and *anatomy*, the study of the human body. Kids enjoy learning about life science because they learn more about their own bodies and about the living world around them. Kids of all ages can have fun watching seeds grow and observing the different behaviors of animals. In life science, the most important topics to learn include the structure and function of living systems, reproduction and heredity of organisms, behavior of organisms, populations and ecosystems, and diversity and adaptation of organisms.

Structure and Function in Living Systems

Living things, such as plants and animals, are called *organisms*. A *system* is defined as the combination of parts of a whole. So organisms, which are made of different parts, can be called *living systems*.

There are millions of different kinds of organisms. The first systematic method of classifying them was invented over 200 years ago by Carolus Linnaeus (1707–1778), a Swedish botanist. This section will begin with a simple model of Linnaeus's way of classifying organisms by structural likeness. The levels of organization of the parts making up organisms will also be studied. The levels of organization of organisms, from the simplest to the most complex, are cells, tissues, organs, and organ systems.

Students will investigate cell structure and the similarities and differences between cells. They will also learn about an organism called a *paramecium* that is made of only one cell. Most organisms are not made of one cell. They are *multicellular* (made of many cells). Students will investigate how cells organize to form organs in multicellular organisms and learn that cells within a multicellular organism are specialized, meaning they have special structures that allow them to have unique functions that benefit the organism. Each kind of cell is necessary for the whole system (organism) to work properly.

Groups

Benchmarks

By the end of grade 5, students should know that

- Organisms are living things.
- A great variety of organisms can be sorted into groups in many ways using various features to decide which organisms belong to which group.
- Features used for grouping depend on the purpose of the grouping.

By the end of grade 8, students should know that

- In classifying organisms, biologists consider details of internal and external structures to be more important than behavior or general appearance.

In this investigation, students are expected to

- Prepare a simple classification system.

Preparing for the Investigation

You can cut the cards ahead of time and put a set for each student or group in a resealable plastic bag. You should also make a copy of the card classification flowchart on page 61 for each student or group.

Presenting the Investigation

1. Introduce the new science terms:

   characteristics Natural features.

   classification The arranging of organisms into groups based on the similarities of their characteristics.

   organism A living thing.

   species A group of similar organisms that can produce more of their own kind.

2. Explore the new science terms:
   - Characteristics are natural similarities, such as structure, development, and biochemical or physiological functions.
   - Biological classification is the arrangement of organisms into categories based on characteristics. The branch of biology

dealing with classification is called *taxonomy* or *systematic classification*.

- The first scheme for classifying animals into groups may have been proposed by Aristotle (384–322 B.C.) more than 2,000 years ago. Since his time, many classification systems have been proposed. While no system is perfect, the Linnaean classification system was found to be the most convenient and is the one used throughout the world today.
- In 1735, Swedish botanist Carolus Linnaeus invented a method for classifying organisms based on characteristics. In the Linnaean classification system, organisms are divided into seven different groups, with each group becoming more specific and containing fewer organisms. The seven groups, from largest to smallest, are kingdom, phylum, class, order, family, genus, and species.

Did You Know?

Carolus Linnaeus (1707–1778) used Latin to name organisms. He even Latinized his own name. His original name was Carl von Linné.

- Classification helps to show how closely related living things are. For example, the table shows the biological classification of a house cat and a

BIOLOGICAL CLASSIFICATION OF HOUSE CAT AND LEOPARD			
Classification	**House Cat**	**Leopard**	**Comparison**
kingdom	Animalia	Animalia	same
phylum	Chordata	Chordata	same
class	Mammalia	Mammalia	same
order	Carnivora	Carnivora	same
family	Felidae	Felidae	same
genus	*Felis*	*Panthera*	different
species	*domesticus*	*pardus*	different

leopard. Notice how many of the classifications of these two animals are the same.

- Organisms in a species naturally reproduce with each other but not with other species.
- While organisms are divided into seven different groups, the two smallest groups, genus and species, are used to name organisms. The first word of a scientific name is the genus and is capitalized. The second word is the species and is not capitalized. This is called *binomial nomenclature*.
- The genus and species name are enough to identify an organism. The table shows the scientific and common names of common organisms.

E X T E N S I O N S

1. Ask students to devise a different method of classifying the cards.

2. Ask students to give the binomial nomenclature of each card.

3. Another way to teach classification to younger students is to use toy animals. Plastic animals can easily be cleaned after handling. Provide the animals or have your students bring them. Lay a bed sheet or a large sheet of paper on the floor and have the students gather around it. Place all the animals in the center of the sheet. Ask for ideas on how to classify the animals. Continue classifying until each animal is in a group by itself.

SCIENTIFIC AND COMMON NAMES OF COMMON ORGANISMS	
Scientific Name	**Common Name**
Rattus norvegicus	common rat or brown rat
Homo sapiens	human
Felis domesticus	house cat
Camelus bactrianus	Bactrian camel
Elephas maximus	Indian elephant
Equus zebra	mountain zebra
Canis lupus	gray wolf
Ulmus americana	American elm
Pinus ponderosa	ponderosa pine

Groups

PURPOSE

To prepare a classification system for different cards.

Materials

scissors
8 unruled index cards—4 yellow, 4 blue
2 crayons—1 blue, 1 yellow
Card Classification flowchart

Procedure

1. Cut the index cards into the following shapes:

   - 2 equal-size large rectangles—1 yellow, 1 blue

   - 2 equal-size small rectangles—1 yellow, 1 blue

   - 2 equal-size large triangles—1 yellow, 1 blue

   - 2 equal-size small triangles—1 yellow, 1 blue

2. Randomly distribute the cards on your desk. Using the crayons, make drawings in circle 1 of the card classification flowchart to represent all the cards.

3. Divide the cards into two groups, using one of these characteristics: shape, color, or size. Label circles 2 and 3 on the flowchart. For example, if you divide the cards into shapes, label the circles on the flowchart "Triangle" and "Rectangle." Make drawings in circles 2 and 3 to represent the two different groups of cards.

4. Repeat step 3, dividing each of circles 2 and 3 into two groups using one of the remaining characteristics in each. Make drawings in each of the four circles (4 to 7) to represent these four groups.

5. Repeat step 3, dividing each of circles 4 to 7 into two groups using one of the remaining characteristics in each. Make drawings in each of the eight circles (8 to 15) to represent these eight groups.

Results

You have created a classification system for the cards.

Why?

You divided the cards into groups by **characteristics** (natural features). Each time the groups are divided, the number of cards in each group gets smaller until there is only one card in each group. The grouping of **organisms** (living things) by characteristics is called **classification.** In this investigation, you started out with eight cards of different color, shape, and size. With each division, the number of cards in each group became less and the similarities between the cards in each group increased. Carolus Linnaeus (1707–1778), a Swedish botanist, invented a method for classifying all organisms based on characteristics. In the Linnaean classification system, organisms are divided into seven different groups, with each group becoming more alike in characteristics and containing fewer organisms. The seven groups, from largest to smallest, are kingdom, phylum, class, order, family, genus, and species. The smallest group, **species,** are similar organisms that naturally produce more of their own kind.

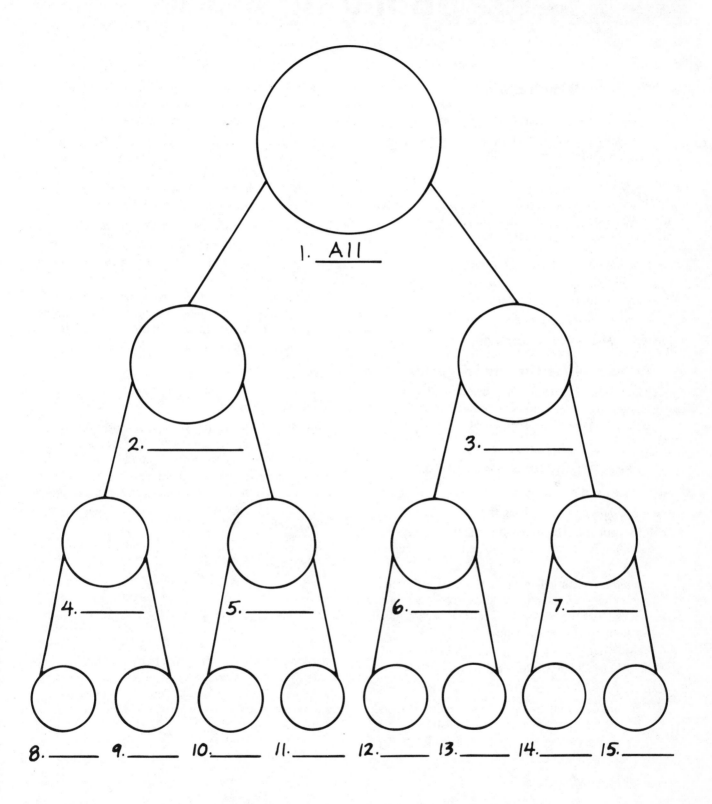

1. __All__

2. _____

3. _____

4. _____

5. _____

6. _____

7. _____

8. ____

9. ____

10. ____

11. ____

12. ____

13. ____

14. ____

15. ____

Building Blocks

Benchmarks

By the end of the grade 5, students should know that

• Some organisms' cells vary greatly in appearance and perform very different roles in the organism.

By the end of grade 8, students should know that

• All organisms are composed of cells, from a single cell to many millions, whose components usually are visible only through a microscope.

In this investigation, students are expected to

• Understand that all organisms are composed of cells that carry on functions to sustain life.
• Represent a cell using a model.
• Identify the basic parts of a cell.

Preparing for the Investigation

Gelatin can be mixed and put in bags ahead of time. The investigation can be a homework assignment if parents are informed that they are to assist with making the gelatin.

Presenting the Investigation

1. Introduce the new science terms:

    cell A building block of living things.

    cell membrane The thin outer skin that holds a cell together and allows materials to move into and out of the cell.

    cytoplasm A clear jellylike material occupying the region between the nucleus and the cell membrane of a cell that contains substances and particles that work together to sustain life.

    nucleus (plural **nuclei**) A spherical or oval-shaped body in a cell that controls cell activity.

2. Explore the new science terms:

    • Cells are generally too small to see with the naked eye. Microscopes are used to see cells.

• Most cells have three basic parts: a cell membrane, a nucleus, and cytoplasm. (A bacterium is one type of cell that does not have a nucleus.)

• Strictly speaking, the cytoplasm is everything between the nucleus and the cell membrane, but often the term is used to indicate only the jellylike material in which cell parts float. This jellylike material is technically called the *cytosol*.

• About two-thirds of a cell's weight is water.

• The cell membrane gives the cell shape, keeps cell parts inside, and regulates the passage of materials into and out of the cell.

• The nucleus provides instructions via chemical composition that guide the life process of every cell. The instructions from the nucleus determine the cell type, for example, a tooth cell or a nerve cell.

Did You Know?

• In 1665, while using a crude microscope, English scientist Robert Hooke (1635–1703) saw tiny spaces in cork. These spaces reminded him of the cells in a monastery, so he called the tiny cavities "cells." The cells viewed by Hooke were dry and empty of protoplasm, which is a term that refers generally to the living parts of a cell.

• The ostrich egg is the largest single cell that has a single nucleus.

E X T E N S I O N

While all animals' cells are basically the same, plant cells and animal cells differ. As a home project, ask some students to make models of animal cells using the basic cell model from "Building Blocks" and adding materials to represent other parts found in animal cells. Ask other students to make models of plant cells.

Building Blocks

PURPOSE

To construct a model that shows the three basic parts of a cell.

Materials

package of lemon gelatin dessert mix
resealable plastic sandwich bag
large grape
adult helper

Procedure

1. With adult assistance, make the lemon gelatin following the package directions. When the gelatin reaches room temperature, put it into the sandwich bag. Seal the bag and put it in the refrigerator for 3 to 4 hours.

2. When the gelatin is firm, open the bag and, using your finger, insert the grape into the center of the gelatin.

Results

You have made a model of a basic cell.

Why?

A **cell** is the smallest building block of living things. Most cells have a cell membrane, a nucleus, and cytoplasm. The **cell membrane,** represented by the plastic bag, is the thin outer skin that holds the cell together and allows materials to move into and out of the cell. The **nucleus,** represented by the grape, is a spherical or oval-shaped body, usually in the center of the cell, that controls cell activity. The **cytoplasm,** represented by the gelatin, occupies the entire region between the nucleus and the cell membrane. The cytoplasm is made of a clear, jellylike material in which the cell parts float.

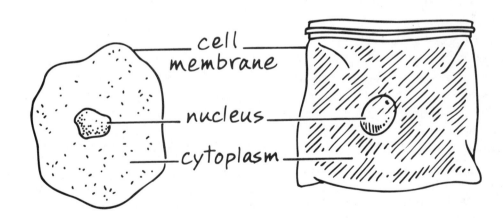

cell membrane

nucleus

cytoplasm

Colored Part

Benchmarks

By the end of grade 5, students should know that

- Plants are made of parts that perform different jobs necessary for life.

By the end of grade 8, students should know that

- One of the most general distinctions between plants and animals is that some plants use sunlight to make their own food.

In this investigation, students are expected to

- Separate chlorophyll from a plant leaf.

Preparing for the Investigation

Collect leaves from the same type of plant for each group. If rocks are not available, other objects of comparable size will work, such as a small piece of wood. Alternatively, students can rub a pencil lead back and forth across the paper instead of striking with a rock.

Presenting the Investigation

1. Introduce the new science terms:

   absorb To take in.

   chlorophyll A green pigment in plant cells that absorbs the light needed for photosynthesis.

   photosynthesis The process by which green plants use chemicals (water and carbon dioxide) in the presence of chlorophyll and light to produce food.

2. Explore the new science terms:
   - Pigments in living organisms are complex chemicals that produce their color by absorbing certain colors of light and reflecting and/or transmitting others. The color that is *reflected* (bounced back) and/or *transmitted* (passed through) is the color you see.
   - Plants are green because of the presence of chlorophyll.
   - Almost all plants contain chlorophyll.
   - Chlorophyll, in the presence of light, uses carbon dioxide and water to produce sugar and oxygen. This is the process of photosynthesis.

- Photosynthesis is something that plants can do but animals cannot do. It is one of the main differences between animals and plants.
- Plants, like most organisms, need water, food, and air to survive.

Did You Know?

- As trees prepare for winter, many of the molecules in the leaves, including chlorophyll, break down and are recycled. This means the atoms of a chlorophyll molecule are used to make other kinds of molecules. With the green chlorophyll gone, the yellow and orange *carotene* pigment in the leaves can be seen.
- The red and purple colors in autumn leaves are due to the production of *anthocyanin*. Cool temperatures increase the sugar content in leaves. A high sugar content and energy from the Sun favor the formation of anthocyanin. Thus, following a period of bright autumn days and cool nights, anthocyanins are produced and the leaves are gloriously colored with red and purple hues.

EXTENSION

Have students try this method of separating different pigments in leaves. Fold the prepared chlorophyll-stained filter in half twice and secure three of the layers with a paper clip, then open to form a cone. Fill a saucer with rubbing alcohol and set the rounded edge of the paper cone in the alcohol. Cover the cone with something, such as an empty plastic soda bottle from which the bottom has been removed, to prevent the alcohol from evaporating. Allow the paper cone to sit undisturbed for 30 minutes or more. Then remove the paper and allow it to dry, which will take about 3 to 5 minutes. Observe the paper to determine if these pigments were present in the leaves: anthocyanin (blue to red), carotene (orange-yellow to red), chlorophyll (blue-green to bright green). *Caution: Keep alcohol away from your eyes, nose, and mouth. Alcohol is flammable, so keep it away from flames.*

Colored Part

PURPOSE

To collect plant pigment from grass.

Materials

3 to 4 sheets of newspaper
6 green leaves
basket-type coffee filter
plum-size rock

Procedure

1. Place the newspaper on a table and stack three of the leaves on top of each other on the newspaper.

2. Cover the leaves with the coffee filter.

3. Strike the filter paper with the rock 15 or more times until a colored area appears on the paper. Be careful not to tear the paper.

4. Allow the colored area to dry, then repeat step 3, using the same filter paper but the other three leaves. You want to collect enough coloring from the leaves to have a dark-colored area on the paper.

Results

A dark green area appears on the paper.

Why?

The color of a leaf is due to the presence of pigments, which are substances that absorb, reflect, and transmit visible light. The green color collected on the filter paper is **chlorophyll,** a green pigment in plant cells. Chlorophyll **absorbs** (takes in) the light needed for **photosynthesis,** the process by which green plants use chemicals (water and carbon dioxide) in the presence of chlorophyll and light to produce food.

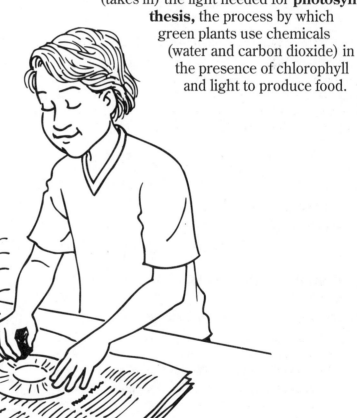

Benchmarks

By the end of grade 5, students should know that

- Some organisms' cells vary greatly in appearance and perform very different roles in the organism, such as those in blood and blood vessels.

By the end of grade 8, students should know that

- The circulatory system moves substances to and from cells where they are needed or produced.

In this investigation, students are expected to

- observe and identify blood vessels of the circulatory system.

Preparing for the Investigation

You will need to provide a small flashlight or penlight and a hand mirror for each student or group.

Presenting the Investigation

1. Introduce the new science terms:

   arteries Large blood vessels that carry red oxygen-rich blood from the heart.

   arterioles Ends of arteries that connect to capillaries.

   blood A fluid that carries materials throughout the body.

   capillaries Microscopic blood vessels that link arteries and veins.

   capillary links A name used for arterioles and venules.

   circulatory system A closed network of blood vessels through which blood flows in the body.

   veins Large blood vessels that carry blue oxygen-poor blood to the heart.

   venules Ends of veins that connect to capillaries.

2. Explore the new science terms:

   - The function of blood is to carry nutrients and oxygen to other body cells and to carry away *waste* (unwanted material), such as carbon dioxide.

Did You Know?

Capillaries are so small that red blood cells must move through them in single file. Because of their small size, there about 1,000 miles of capillaries per square inch (256 km/cm²) of the body, or enough capillaries in the adult body to circle Earth.

   - Blood is a liquid tissue. The liquid in which blood cells float, which is made mostly of water, is called *plasma*.

   - In the body's circulatory system, blood moves from the heart through blood vessels called arteries and capillaries and then back to the heart through veins. The blood carries nutrients to cells and carries away waste.

EXTENSION

Have your students help prepare a display representing the amount of blood in a baby, in a child, and in an adult. Fill nine 1-quart (1-liter) jars with water. Add 10 drops of red food coloring to each jar and stir. Fold 3 index cards in half so that they stand up. Draw a baby and a jar filled with red liquid on one card. Label the card "Baby." Stand this card in front of one of the jars. On the second card, draw a child and three jars filled with red liquid. Label the card "Child" and stand the card in front of three of the jars. On the third card, draw an adult and five jars filled with red liquid. Label the card "Adult" and stand the card in front of the remaining five jars.

Connectors

PURPOSE

To identify blood vessels.

Materials

flashlight
hand mirror

Procedure

1. Raise your tongue and shine the light on the area under the tongue.
2. Use the mirror to inspect the area under your tongue.
3. Find the parts identified in the drawing.

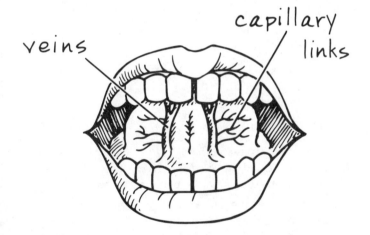

Results

Large blue blood vessels and hair-thin red and bluish-red blood vessels are seen under the tongue.

Why?

Blood is a fluid that carries materials throughout the body. Blood flows through a closed network of blood vessels called the **circulatory system.** The main blood vessels in this system are arteries, veins, and capillaries. Veins are large blood vessels carrying blue oxygen-poor blood to the heart. **Veins** are often seen beneath the skin. **Arteries** are also large vessels, but are generally not near the skin, so they are not seen. Arteries carry red oxygen-rich blood from the heart. **Capillaries** are microscopic blood vessels that link arteries and veins. Capillaries connected to veins contain bluish-red oxygen-poor blood. The ends of veins and arteries that connect to capillaries are called **venules** and **arterioles,** respectively. Venules and arterioles can be called **capillary links.** Veins and capillary links can be seen under the tongue.

Slipper Animal

Benchmarks

By the end of grade 5, students should know that
- Some living things consist of a single cell.

By the end of grade 8, students should know that
- Some kinds of organisms, many of them microscopic, cannot be neatly classified as either plants or animals.

In this investigation, students are expected to
- Represent a paramecium using a model.

Preparing for the Investigation

Note that the color of the paper is not significant. It does not represent the color of paramecia, which are basically colorless and transparent.

Presenting the Investigation

1. Introduce the new science terms:

   cilia Tiny hairlike parts used by some unicellular organisms for locomotion.

   locomotion The act of moving from one place to another.

   paramecium (plural **paramecia**) A protist that has cilia and two kinds of nuclei.

   protist An organism of the kingdom Protista, which includes most of the unicellular organisms having visible nuclei.

   reproduction The process by which an organism produces young of the same species.

   unicellular One-celled.

2. Explore the new science terms:
   - Some biologists call the Protista the *Protoctista*.
   - Most protists can move about in search of food or light.
   - A paramecium is a protist that uses cilia for locomotion.
   - Protists live in water.
   - Natural sources of freshwater, such as ponds and lakes, contain protists, even if the water looks crystal clear. A paramecium is a protist found in freshwater.
   - A paramecium has two kinds of nuclei, *macronuclei* (large) and *micronuclei* (small). Micronuclei control reproduction, and macronuclei control normal cell activities.

Did You Know?

In 1870, Louis Pasteur (1822–1895), a French chemist and microbiologist, saved the silk industry by identifying a protist that caused a disease of silkworms.

EXTENSION

1. Have students use a biology text to identify the shape and function of other body parts of paramecia. These parts can be drawn on the paramecium model.

2. Have the class use a microscope to look at prepared slides of paramecia or prepare the slides themselves. (See appendix 3 for a list of science supply companies where prepared slides and live specimens can be purchased.)

Slipper Animal

PURPOSE

To make a model of a paramecium.

Materials

2 left shoes—1 large, 1 small
2 sheets of construction paper—1 light, 1 dark
pencil
scissors
glue
glitter—2 different colors
marking pen

Procedure

1. Set the large shoe on the light-colored paper and the small shoe on the dark-colored paper.

2. Draw around the sole of each shoe, then cut out the sole shapes.

3. Cover the edge of the underside of the small paper sole with glue.

4. Glue the small sole to the middle of the large sole.

5. Cut slits in the large sole to the edge of the small sole to make a fringe around the entire edge of the light-colored paper.

6. Place a large spot of glue in the center of the small sole. Cover the spot of glue with one of the colors of glitter.

7. When the glue dries, place another small spot of glue on one side of the large spot of glitter and cover it with the other color of glitter.

Paramecium

Results

A model of a paramecium is made.

Why?

A **paramecium** is a **unicellular** (one-celled) organism that has two kinds of nuclei. Most unicellular organisms having visible nuclei are **protists,** grouped in a kingdom called Protista. Paramecia are shoe-shaped, as in the model in this investigation. The fringe around the edge of the model paramecium represents tiny hairlike parts of the cell called **cilia,** which are used for **locomotion** (the act of moving from one place to another). The two spots of glitter represent the two kinds of nuclei of the paramecium. The larger nucleus controls normal cell activities, while the smaller nucleus controls **reproduction** (the process by which an organism such as the paramecium produces young of the same species).

Reproduction and Heredity

In this section, *reproduction*, the process by which living things produce more of their own kind, will be studied. Reproduction can be *asexual* (having one parent) or *sexual* (having two parents).

One of the most interesting aspects of sexual reproduction is *heredity*, how physical characteristics are passed down from one generation to the next. A child may be said to have his mother's nose and his father's eyes. Another child in the family may not resemble either parent, but is said to look very much like a great-grandparent. Some parents have brown hair and their child has red. By investigating heredity in this section, students will discover how these things can happen.

Characteristics

Benchmarks

By the end of grade 5, students should know that

- For offspring to resemble their parents, there must be a way that information is transferred from one generation to the next.

By the end of grade 8, students should know that

- In sexual reproduction, typically half of the genes in the offspring come from each parent.

In this investigation, students are expected to

- Identify some inherited traits of humans.

Preparing for the Investigation

This investigation could be a homework project. If it is done in class, students within a lab group can share a mirror.

Presenting the Investigation

1. Introduce the new science terms:

    heredity The passing of traits from one generation to the next.

    inherit To receive traits from parents.

    phenotype A description of a trait.

    trait A physical characteristic.

2. Explore the new science terms:
    - Eye color is a physical characteristic, called a trait, and brown eyes is an aspect of physical appearance, called a phenotype.
    - A boy with brown eyes inherited his eye color from his parents or grandparents or other ancestors. The passing of the trait of eye color from parent to child is called heredity.

Did You Know?

"Identical twins" are not 100 percent identical. While they may look alike, there are some differences of phenotype, such as their fingerprints. The fingerprint patterns for identical twins are the same, but there are differences, such as the number of ridges.

EXTENSION

Make a data table on the chalkboard for the traits of everyone in the class. In column 3, instead of "Your Phenotype" put the heading "Number." Enter the number of students having each trait in the table to determine which visible characteristics are represented by a greater number of students class.

26 Characteristics

PURPOSE
To identify visible traits.

Materials

mirror
pencil

Procedure

1. Using the mirror, observe the color of your eyes. Record your eye color as brown or other in the Your Phenotype column of the Individual Traits Data table.

2. Use the mirror to study your tongue. Try to roll your tongue as shown. Record in the table whether you are a roller or a nonroller.

3. Use the mirror to study your ears. Compare your earlobes to the ones shown to determine whether you have attached or unattached earlobes. Record your earlobe type in the table.

4. Determine whether you are right- or left-handed by identifying the hand you write with. Record your handedness in the table.

Results

You created a chart of your observable characteristics.

Why?

Eye color, ability to roll your tongue, attachment of earlobes, and right- or left-handedness are all physical characteristics called **traits.** A description of traits, such as brown eyes, is a **phenotype.** The passing of these traits from one generation to the next is called **heredity.** You have **inherited** each of these traits, which means you have received them from your parents. Phenotypes are expressions of specific traits. For example, eye color is a trait, and the actual color of your eyes is your phenotype for this trait.

Earlobe

Attached   Unattached

Tongue

Roller   Non roller

INDIVIDUAL TRAITS DATA		
Trait	Phenotype	Your Phenotye
eye color	brown other	
tongue rolling	roller nonroller	
earlobes	attached unattached	
handedness	right-handed left-handed	

Matches

Benchmarks

By the end of grade 5, students should know that
- Some likenesses between children and parents, such as eye color, are inherited.

By the end of grade 8, students should know that
- In sexual reproduction, typically half of the genes in the offspring come from each parent.

In this investigation, students are expected to
- Identify chromosome pairs.
- Explain the role of genes in inheritance.

Preparing for the Investigation

You will need to make a copy of the Chromosome Models—A sheet (on page 75) for each student or group.

Presenting the Investigation

1. Introduce the new science terms:

   allele One of several different forms of a specific gene.

   chromosome A rod-shaped structure in the nucleus of a cell that contains DNA.

   deoxyribonucleic acid (DNA) Chemical molecules in chromosomes that control cell activity and determine hereditary traits.

   gene The part of a chromosome that determines hereditary traits; consists of DNA.

   gene site The site where a gene is positioned on a chromosome.

2. Explore the new science terms:
   - Traits are determined by genes, and all genes consist of DNA.
   - The gene site on the chromosome for a trait such as a flower's color is always the same for a type of flower. Gene site is also called *locus*.
   - Flowers have different colors, so there are different forms of genes for the trait of flower color. These alternative forms are called *alleles*.
   - Each living cell has chromosome pairs, and each chromosome in the pair has a copy of each gene. (The exception is sex cells, which have single chromosomes.)

- Although many alleles may exist for a given gene, each organism has only two alleles for that gene, and they may or may not be the same.
- The main function of a gene is to control the production of a substance called protein. The kind and number of proteins determine the trait of an organism.

Did You Know?

The first person known to discover the basic laws of heredity and suggest the existence of genes was an Austrian monk, Gregor Johann Mendel (1822–1884). Mendel published his findings in 1866, but his work went unnoticed until 1900, when three scientists in three different countries almost simultaneously rediscovered Mendel's work.

EXTENSION

Each form of the genes (alleles) in each chromosome pair in the investigation are identical, but in most real chromosome pairs, the paired alleles are not all identical. For example, the allele for blue eyes might be on one chromosome and the allele for brown eyes on the matching chromosome. Make a copy of the Chromosome Models—B sheet on page 76 for each student or group. You may wish to make these copies on colored paper so as not to confuse them with the original models. Point out that each gene type is identified by size and that different alleles for the same trait are indicated by open and shaded or patterned areas in the models.

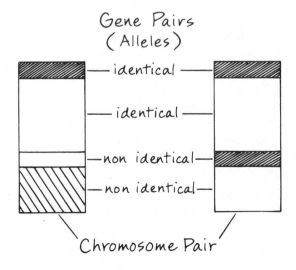

Gene Pairs (Alleles)
— identical —
— identical —
— non identical —
— non identical —
Chromosome Pair

Matches

PURPOSE

To show how identical genes on paired chromosomes match.

Materials

scissors
copy of the Chromosome Models—A sheet

Procedure

1. Cut out each chromosome model.
2. Spread the chromosome models on your desk or lab table.
3. Turn the models as needed to pair up as many matching models as possible. A chromosome pair with identical genes must match in length as well as in the number, size, and location of open and shaded areas, as shown in the diagram.

Results

In the diagram, four chromosome pairs are identified.

Why?

Chromosomes are rod-shaped structures in the nucleus of a cell that contain **genes,** which are made of chemical molecules called **deoxyribonucleic acid (DNA).** DNA controls cell activity and determines hereditary traits. The different open and shaded areas on the chromosome models represent **gene sites** (sites where genes are located on a chromosome). Each living cell contains chromosome pairs, and each pair has matching gene sites. Each member of a gene pair is called an **allele,** which is one of several different forms of a specific gene.

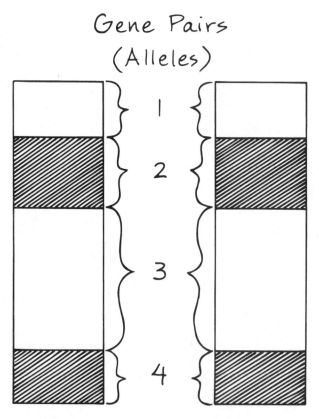

Gene Pairs
(Alleles)

Chromosome Pair

Chromosome Models – A

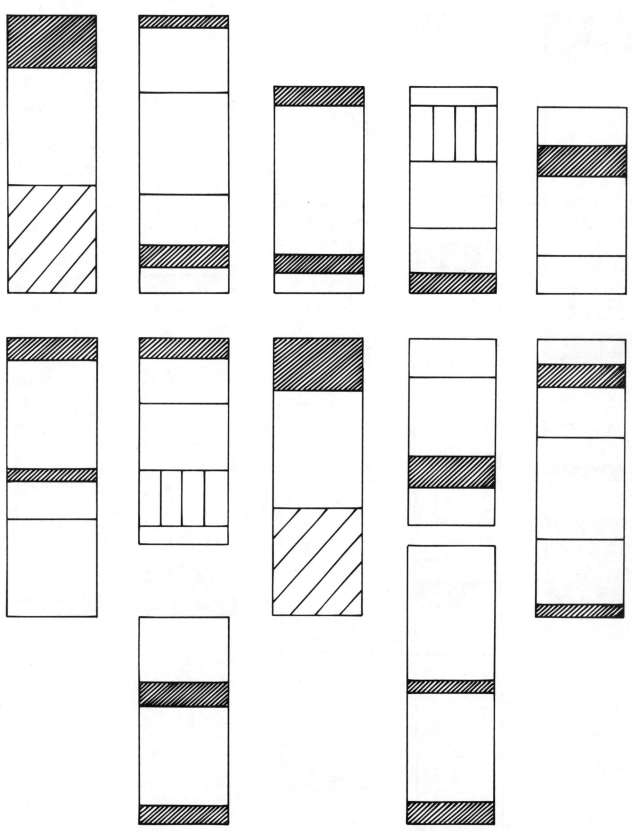

Chromosome Models - B

One to Four

Benchmarks

By the end of grade 5, students should know that

- For offspring to resemble their parents, there must be a way that information is transferred from one generation to the next.

By the end of grade 8, students should know that

- Animals have a number of body structures that contribute to their being able to reproduce.
- In sexual reproduction, a single specialized cell from a female merges with a specialized cell from a male.

In this investigation, students are expected to

- Model meiosis, the production of sex cells.
- Distinguish between sex cells and other cells.

Preparing for the Investigation

The poster board can be colored or white, or you can use bulletin board paper of comparable size. The large lid should be about 6 inches (15 cm) in diameter, the small lid about 4 inches (10 cm) in diameter. Cans or bowls can be used in place of the lids.

Presenting the Investigation

1. Introduce the new science terms:

   egg A female sex cell.

   gender The sex of an organism: male or female.

   meiosis The process of cell division by which sex cells are produced.

   sex cells Specialized cells, sperm and eggs, produced by meiosis.

   sex chromosome A chromosome that contains the gene for gender and is known as an X or Y chromosome.

   sperm A male sex cell.

2. Explore the new science terms:
   - Most human cells have 46 chromosomes, or 23 pairs of chromosomes from the union of sperm and egg.
   - During the process of meiosis, a cell divides two times, forming four sex cells.
   - As a result of meiosis, each new human sex cell has 23 chromosomes, which is half the number of chromosomes that are present in other body cells. One of the 23 chromosomes is a sex chromosome, X or Y. Point out that in the investigation, only the X chromosomes are represented.
   - After union of the two human sex cells, the resulting cell contains 23 pairs of chromosomes.
   - Eggs have one X chromosome.
   - Sperm have either an X or a Y chromosome.
   - Female cells have a pair of X chromosomes (XX).
   - Male cells have one X chromosome and one Y chromosome (XY). These chromosomes, while called "pairs," do not match as do other chromosome pairs. The X chromosome is longer than the Y chromosome.

Did You Know?

There are hundreds of genes on the X chromosome that are not found on the smaller Y chromosome. These are called *sex-linked genes*. Hemophilia, or "bleeder's disease," is an example of a sex-linked disease.

EXTENSION

Repeat the investigation, using models of male chromosomes to demonstrate how sperm are produced. The difference will be in the sex chromosomes represented by the two circles cut from the index card. Write an X on one circle and a Y on the other circle. Ask the students the following questions about the results of this investigation.

QUESTIONS FOR STUDENTS		
Questions	female cell	male cell
1. How many total chromosomes are in cell A before duplication?	6	6
2. How many matching pairs of chromosomes are in cell A?	3	2
3. How many unmatched chromsomes are in cell A?	0	1
4. After duplication, how many chromosomes are there in cell A?	12	12
5. How many chromosomes are in cell B?	6	6
6. How many chromosomes are in cell C?	6	6
7. How many total chromosomes are in each sex cell?	3	3
8. How does the number of chromosomes in each sex cell compare to the number that was in cell A before duplication occurred?	half as many	half as many
9. How do the types of chromosomes in each sex cell compare?	same in each	2 sets of similar cells
10. Which sex cells are present?	XX	XY

One to Four

PURPOSE

To model meiosis.

Materials

marker
2 lids—1 large, 1 small
22-by-28-inch (55-by-70-cm) sheet of poster board
2 unruled index cards—two different colors
pen
scissors

Procedure

1. Use the marker and the lids to draw seven circles on the poster board. Draw arrows connecting the circles and label the circles as shown.

2. Fold one of the index cards in half, long sides together.

3. Using the pen, draw the largest possible circle, triangle, and square on one side of the folded card. Cut these shapes out, cutting through both layers of paper. You will have two cutouts of each shape.

4. Repeat steps 2 and 3, using the other index card.

5. Write an X on each cutout circle.

6. Randomly place one shape of each color in cell A, the body cell, on the poster board.

7. Place the remaining shapes in cell A, stacking shapes of the same color. Each colored shape has been duplicated, forming a stacked pair.

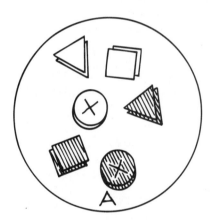

8. Move one stacked pair of each shape into cell B. The color of the shapes doesn't have to be the same.

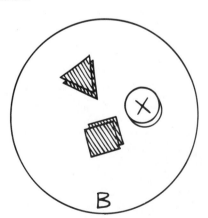

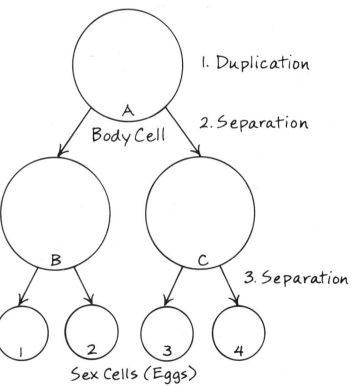

9. Move the remaining shapes from cell A into cell C.

10. Separate one of the pairs of shapes in cell B, placing one shape in sex cell 1 and the other in sex cell 2. Repeat with the remaining two pairs of shapes.

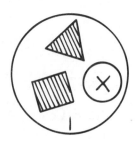

11. Separate one of the pairs of shapes in cell C, placing one shape in sex cell 3 and the other in sex cell 4. Repeat with the remaining two pairs of shapes.

Results

One body cell with a double set of three shapes is first duplicated, then goes through divisions, forming four sex cells with one of each shape in each.

Why?

Sex cells, called **eggs** and **sperm** from the female and the male, respectively, are specialized cells that each contain one chromosome from each parent. One of these chromosomes is a **sex chromosome,** which contains the gene for **gender** (sex of organism, male or female) and is known as the X or Y chromosome. Sex cells are produced by a process of cell division called **meiosis.** During meiosis, there are many steps. In this investigation, the process has been simplified to show only three steps. Before meiosis starts, the body cell has a double set of chromosomes, one from each parent. In this investigation, one member of the set has the same shape as the other, but is represented by a different color. The first step is the duplication of the chromosomes (placing two of each colored shape in the body cell). In the second step, the single cell divides, forming two cells (B and C), each with a random combination of chromosomes from the parents (represented by color), but with one pair of each type of chromosome (shape). In the third step, cells B and C split in half, forming four sex cells. The chromosome pairs in cells B and C split and one chromosome from each pair is in a sex cell. Four cells form from the original single cell. These cells are called the sex cells, and each has half the number of chromosomes as the original body cell, but each has one of each kind of chromosome found in the body cell (circle, triangle, square). In females, the body cell contains a pair of X chromosomes and forms four egg cells that each contain one X chromosome (represented by the circle marked with an X).

Combinations

Benchmarks

By the end of grade 5, students should know that

- Some likenesses between children and parents, such as eye color, are inherited.

By the end of grade 8, students should know that

- In sexual reproduction, typically half of the genes in the offspring come from each parent.

In this investigation, students are expected to

- Distinguish between dominant and recessive traits and understand that inherited traits of an individual are contained in genetic material.

Preparing for the Investigation

Bulletin board paper of comparable size can be used instead of poster board.

Presenting the Investigation

1. Introduce the new science terms:

   dominant allele A gene form that when present determines the trait.

   fertilization The joining of two sex cells, an egg and a sperm, from two parents.

   genotype The genetic makeup of an organism or a group of organisms as determined by alleles.

   phenotype The observable characteristics of an organism that are determined by genotype; the expression of specific traits.

   recessive allele A gene form that does not determine a trait when a dominant allele is present.

   sexual reproduction Reproduction by fertilization.

   zygote A cell formed by the joining of a sperm and an egg.

2. Explore the new science terms:

   - In humans, fertilization results in the union of a human egg with 23 chromosomes and a sperm with 23 chromosomes, producing a zygote with 23 chromosome pairs.

   - Each zygote has one allele from each parent for each kind of trait. So there are two alleles for each trait. When a dominant and a recessive allele are present, the dominant allele determines the trait.

   - For a dominant allele, a capital letter is used, such as D for dimpled chin.

   - For a recessive allele, a lowercase letter is used, such as d for no dimpled chin. Any letter can be used, but only one letter is used for each trait. The letter for the dominant allele is always written first.

   - Because some alleles are dominant, an organism's appearance does not reflect its genetic makeup. So scientists distinguish between an organism's visible characteristics, called its phenotype, and its genetic makeup, its genotype.

Did You Know?

It is possible for two brown-eyed parents to have a blue-eyed child with the genotype of Bb. This is because the presence of other genes masks the dominant allele B, so that the brown pigment would not be produced. About 1 in every 50 persons with a genotype of Bb will have blue eyes.

EXTENSIONS

1. You may wish to make sets of plant and animal cards. One way to distinguish the animal cards from the plant cards would be to use ruled cards for plants and unruled cards for animals.

2. Many cells form from the zygote by a process called *mitosis*. Research can be done on this type of cell division. Dogs, cats, and humans all started out as zygotes.

Combinations

PURPOSE

To use allele pairs to determine the characteristics of a person.

Materials

marker
14 colored index cards—7 each of 2 colors
one 22-by-28-inch (55-by-70-cm) piece of poster board

Procedure

1. Use the marker to draw a symbol for female on one side of all the cards of one color and a symbol for male on one side of all the cards of the other color.

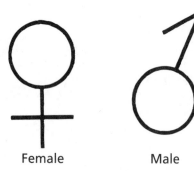

Female   Male

2. From the list, randomly select a dominant or recessive allele for each of the traits and write the letter for each trait on the back of the cards of one color. Repeat with the cards of the other color. You can use a dominant allele twice and not use the recessive allele for that trait or vise versa. Each card represents one member of a gene pair.

3. Use the marker to draw a circle as large as possible on the poster board.

4. Place all the female and male cards in the circle, allele side up, so that they do not overlap.

5. Pair the female and the male cards by traits so that there are seven pairs of cards. Stack the cards so that each chromosome pair consists of two cards that relate to the same trait. For example, if the female card has a D or a d, then the male card must have a D or a d also.

6. Record the letter combination of the alleles for each chromosome pair in the "Genotype" column of the Traits Data table. Write the capital letter, if any, first. For example, if the cards show a D and a d, write "Dd."

7. Record the expressed trait of each genotype in the "Phenotype" column of the table. For example, if the genotype is DD or Dd, write "dimpled chin." If the genotype is dd, write "no dimpled chin." Note that only when the genotype has a dominant allele, expressed by a capital letter, is the trait expressed.

DOMINANT AND RECESSIVE ALLELES FOR TRAITS		
Trait	Dominant Allele	Recessive Allele
earlobe attachment	unattached earlobe (L)	attached earlobe (l)
hair color	brown hair (R)	red hair (r)
hair shape	widow's peak (W)	no widow's peak (w)
ability to roll tongue	tongue rolling (T)	non–tongue rolling (t)
handedness	right-handedness (H)	left-handedness (h)
eye color	brown eyes (B)	blue eyes (b)
dimpled chin	dimpled chin (D)	no dimpled chin (d)

8. Based on the data in the "Phenotype" column, describe the appearance of a person with the allele pairs in this investigation.

Results

The description of the person varies with the makeup of the genes.

Why?

Reproduction is the process by which an organism produces young of the same species. Most animals reproduce by **sexual reproduction,** which involves the union of a sex cell, an egg and a sperm, from two parents. The joining of these cells is called **fertilization.** The joining of a sperm and an egg produces a single cell called a **zygote.** In this activity, the zygote is represented by the circle drawn on the paper. The cards represent allele pairs from a female sex cell (egg) and a male sex cell (sperm).

Each trait has two possible alleles—one **dominant allele,** represented by a capital letter, and one **recessive allele,** represented by a lowercase letter. There are two alleles for each trait in the zygote, one from the sperm and one from the egg. The allele pair for each trait is represented by the combination of letters for each allele, with the capital letter, if any, written first. The alleles determine the **genotype** (genetic makeup) of an organism or a group of organisms. The expressed traits, or **phenotype,** of an organism are determined by the genotype. If the genotype for a specific trait has one or two dominant alleles, the trait of the dominant allele is expressed in the organism. For example, an organism with the genotype DD or Dd in the investigation will have a dimpled chin. The recessive trait is not expressed when the dominant allele is present. If only recessive alleles are present in the genotype, such as dd, the recessive trait will be expressed. An organism with the genotype dd will not have a dimpled chin.

TRAITS DATA	
Genotype	**Phenotype**

Teaching Tips for the Investigation

Checkerboard

Benchmarks

By the end of grade 5, students should know that
- Children may look like grandparents instead of parents.

By the end of grade 8, students should know that
- In sexual reproduction, typically half of the genes in the offspring come from each parent.

In this investigation, students are expected to
- Distinguish between dominant and recessive traits and understand that inherited traits of an individual are contained in genetic material.
- Use a Punnett square to determine the possible genotypes of offspring based on various gene combinations.
- Distinguish between purebred and hybrid organisms.

Presenting the Investigation

1. Introduce the new science terms:

   hybrid An offspring whose alleles are different for a trait.

   Punnett square A grid used to determine the percentage of possible genotypes of offspring based on parental gene combinations.

   pure trait An offspring whose alleles are the same for a trait.

2. Explore the new science terms:
   - The genotypes DD and dd are called *purebreds* because both alleles are identical.
   - The genotype Dd has two unlike alleles and is called a *hybrid*. (The letter for the dominant allele is always written first.)
   - The figure shows a Punnett square for mouse parents having the genotypes Bb and Bb (hybrid black). The trait represented is fur color, with the dominant allele being black (B) and the recessive allele being white (b). In the figure, the possible genotypes of offspring are shown. The percentage of occurrence of each genotype can be determined by the number of squares in which the genotype appears, as follows:

1 out of 4 = 25%	3 out of 4 = 75%
2 out of 4 = 50%	4 out of 4 = 100%

Thus, 25 percent of offspring will possibly be of the genotype BB (purebred black), 25 percent bb (purebred white), and 50 percent Bb (hybrid black).

Did You Know?

Before the twentieth century, people thought that traits were passed from one generation to the next through the blood. Today people still speak of "blood relatives."

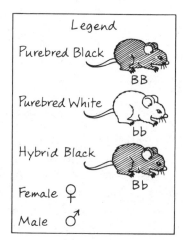

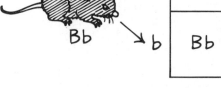

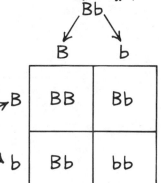

EXTENSION

Have students prepare Punnett squares for other parental gene combinations, such as those in the Parental Gene Combinations table.

PARENTAL GENE COMBINATIONS

Female	Male	Characteristics
Tt	tt	T = tongue roller, t = non–tongue roller
WW	ww	W = widow's peak, w = no widow's peak
hh	Hh	H = right-handed, h = left-handed
Bb	Bb	B = brown eyes, b = blue eyes
DD	DD	D = dimpled chin, d = no dimpled chin

Janice VanCleave's Teaching the Fun of Science

Checkerboard

PURPOSE

To determine the possible genotypes of offspring.

Materials

pencil
ruler
sheet of typing paper

Procedure

1. Use the pencil and ruler to draw on the paper a square containing four boxes.
2. Select one of these parent combinations:
   - Father (RR) + Mother (Rr)
   - Father (Rr) + Mother (Rr)
3. Write the two possible female alleles for the egg along the top edge of the square as shown.
4. Write the two possible male alleles for the sperm down the left edge of the square as shown.

5. Fill in the squares by writing the letter combination for each allele pair in each box. Write the letter for the dominant allele first. For example, the letter combination for the top left box is RR.
6. Use the data in the square to determine the percentage of occurrence of each genotype for offspring that are purebred red (RR), purebred brown (rr), or hybrid red (Rr). To do this, count the number of squares in which each genotype appears. The percentage of occurrence is 25 percent when the genotype is in one square, 50 percent in two squares, 75 percent in three squares, and 100 percent in all four squares.

Results

The genotypes of offspring having different parental gene combinations are determined. For combination RR + Rr, the percentage of occurrence of each genotype is 50 percent RR and 50 percent Rr. For the combination Rr + Rr, the percentage of occurrence of each genotype is 25 percent RR, 50 percent Rr, and 25 percent rr.

Why?

The square used to determine genotypes of offspring of parents is called a **Punnett square.** The more common the combination of alleles in the Punnett square, the more probable that an offspring will have that trait. Offspring with identical alleles for a trait are called **pure traits,** and offspring with different alleles for a trait are called **hybrids.**

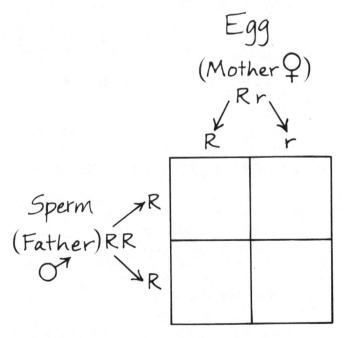

Baby Plant

Benchmarks

By the end of grade 5, students should know that

- Some organisms' cells, such as the different parts of a plant seed, vary greatly in appearance and perform very different roles.

By the end of grade 8, students should know that

- After the union of plant sex cells, the fertilized egg multiplies to form a multicellular seed.

In this investigation, students are expected to

- Observe a seed and identify its parts and their functions.

Preparing for the Investigation

Soak four pinto beans per student or group in water overnight. Keep the beans in the refrigerator to prevent souring. The extra beans are in case of accidental breakage of the beans during dissection.

Presenting the Investigation

1. Introduce the new science terms:

   cotyledon A simple leaf beneath a seed coat that stores food for a developing plant.

   embryo An organism in the earliest stage of its development, such as the immature plant inside a seed.

   epicotyl The part of a plant embryo above the cotyledons' point of attachment that develops into a plant's stem, leaves, flowers, and fruit.

   hypocotyl The part of a plant embryo beneath the cotyledons' point of attachment.

   plumule The tiny, immature leaves located at the tip of an epicotyl that at maturity form the first true leaves of a plant.

   radicle The lower part of a hypocotyl that develops into a plant's root system.

   root system The part of a plant that grows down into the soil, from which it takes in water and nutrients.

   seed A product of sexual reproduction in plants that contains genetic material from both parents and can develop into a mature plant.

   seed coat The protective outer covering of a seed.

   stem The central support structure of a plant.

2. Explore the new science terms:

   - While a cotyledon is also called a *seed leaf*, it doesn't have the appearance of a leaf. As the plant grows and uses the food in the cotyledon, the cotyledon gets smaller and finally dries up and falls off.

   - The embryo of a plant has the appearance of a tiny "baby" plant. The immature plant inside a seed develops into a mature plant.

   - The hypocotyl connects the epicotyl and radicle.

   - During the germination (growth) of beans, the hypocotyl forms a sharp bend or hook that breaks through the soil. When the hook straightens, it lifts the epicotyl out of the ground.

   - A plant's stem supports the leaves and flowers of the plant and transports water, minerals, and food throughout the plant.

Did You Know?

The largest seed of all plants is produced by the coco-de-mer, a palm tree that grows in the Seychelles. The seed can weigh up to 50 pounds (23 kg).

EXTENSIONS

1. Try the investigation using raw peanuts. Students can remove the peanut from its shell and peel away the seed coat. The two cotyledons can be separated and an embryo will be found inside each. Have students compare the embryo of the peanut to that of the bean.

2. Some plants, such as beans, have two cotyledons and are called *dicotyledons*. Other plants, such as corn, have one cotyledon and are called *monocotyledons*. Ask students to find out more about dicotyledons and monocotyledons. How do their leaves and flowers differ? (The leaves of typical monocots have parallel veins and the flower parts are in multiples of three. The leaves of typical dicots are net-veined and the flower parts are in multiples of four or five.)

Baby Plant

PURPOSE

To identify the parts of a seed.

Materials

2 pinto beans soaked in water overnight
paper towel
magnifying lens

Procedure

1. Place the beans on a paper towel to dry.
2. Scratch off the covering of one of the beans with your fingernail.
3. Gently break this bean apart with your fingers and use the magnifying glass to look at the inside of the bean.
4. Use the diagram to identify the following parts of the bean: cotyledon, embryo, epicotyl, hypocotyl, radicle, and plumule.
5. Repeat steps 2 to 4 with the other bean. Compare the insides of each bean.

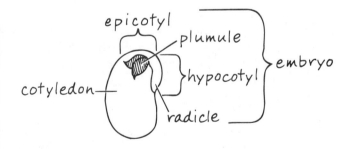

Results

A pinto bean has been dissected and its parts identified. Both beans are the same inside.

Why?

A **seed** is the product of sexual reproduction in plants. A seed contains genetic material from both parents and can develop into a mature plant. A bean is a seed. Beneath the outer protective covering of the seed, called the **seed coat,** are two simple leaves that fit together, called the **cotyledons.** These parts contain the food for the developing plant. Inside and attached to the cotyledons is an immature plant called an **embryo** (an organism in the earliest stage of its development). The part of the embryo beneath the cotyledons' point of attachment is the **hypocotyl.** At the lower end of the hypocotyl is the **radicle,** which develops into a **root system** (part of a plant that grows down into the soil, from which the plant takes in water and nutrients). The part of the embryo above the cotyledons' point of attachment is called the **epicotyl.** This part develops into the plant's **stem** (central support structure of a plant), leaves, flowers, and fruit. The tiny, immature leaves located at the tip of the epicotyl are called the **plumule.** At maturity these leaves form the first true leaves of the plant.

Sprouter

Benchmarks

By the end of grade 5, students should know that

- For offspring to resemble their parents, there must be a way that information is transferred from one generation to the next.

By the end of grade 8, students should know that

- Plants have a variety of body structures that contribute to their being able to reproduce.

In this investigation, students are expected to

- Identify a form of asexual reproduction.
- Understand that asexual reproduction is reproduction from a single parent.

Preparing for the Investigation

You may wish to have students do this as a home project, with a different student from each group using stems from a different plant. Plants that grow well from cuttings include ivy, geraniums, mother-in-law's tongue, and spider plant.

Presenting the Investigation

1. Introduce the new science terms:

   asexual reproduction Reproduction in which there is only one parent and the offspring are identical to the parent.

   cutting A piece cut from a plant to grow a new plant.

   vegetative propagation Production of a new plant from a plant part other than a seed.

2. Explore the new science terms:
   - Growing plants from cuttings is considered an artificial type of vegetative propagation.
   - Examples of natural vegetative propagation are the growing of plants from *bulbs* (short underground stems with thick fleshy leaves that contain stored food) or *tubers* (fleshy underground stems with buds from which new plants develop).
   - Artificial and natural vegetative propagation require only one plant, so they are examples of asexual reproduction.
   - Vegetative propagation can be used to grow plants faster and at times more successfully than with seeds.
   - Plants grown by vegetative propagation are exactly like the parent plant, so they could be called *clones*.

Did You Know?

Seedless fruits, such as seedless oranges and grapes, are produced by vegetative propagation.

EXTENSION

Have your students investigate natural vegetative propagation using bulbs, such as onions, and tubers, such as potatoes. Both of these will develop roots if you put them each in a jar of water. Place the root end of the onion just below the surface of the water. The potato should be placed so that it is about halfway into the water. Stick toothpicks into the onion or potato to support them in the jar if necessary.

Sprouter

PURPOSE

To grow a plant from a cutting.

Materials

1-pint (500-ml) jar
tap water
scissors
leafy stem of an ivy plant

Procedure

1. Fill the jar with water.
2. Use the scissors to cut across the end of the plant stem.
3. Stand the cut end of the stem in the jar of water.
4. Observe the cut end of the stem for 2 or more weeks.

Results

Tiny roots start to grow on the stem.

Why?

Many plants, such as the ivy in this investigation, will easily form roots on a **cutting** (a piece cut from a plant to grow a new plant). With the new roots, the stem can develop into a mature plant. This type of reproduction of a new plant is called **vegetative reproduction** (the production of a new plant from a plant part other than a seed). Vegetative reproduction is an example of **asexual reproduction,** which is reproduction in which there is only one parent and the offspring are identical to the parent.

Behavior

Behavior is a fundamental characteristic of animal life. *Ethology* is the biological study of animal behavior. All behavior is an observable reaction to a *stimulus*, simple or complex. The response to the stimulus can be *innate*, or unchangeable and predictable; or it can be *learned*, or changed. Most behavior is not considered strictly innate or learned, but rather a blend of the two. In this section, students will investigate both of these behavior types.

The more developed the nervous system of an organism, the more varied its behavior. Organisms that do not have nervous systems, such as plants, have limited behavior. Responses from plants do not come from nerve messages. Instead, they are due to physical changes, such as changes in cell size. In this section, students will also investigate different types of plant behavior.

Sniffer

Benchmarks

By the end of grade 5, students should know that
- Human behavior is influenced by the memory of past experiences.

By the end of grade 8, students should know that
- Generally, all behavior is affected by both inheritance and experience.
- Humans can detect a great range of olfactory (smell) stimuli.

In this investigation, students are expected to
- Identify responses in organisms to external stimuli, such as smells.
- Distinguish between innate and learned behavior.

Preparing for the Investigation

Prepare one set of four sniffing bags for each group or pair of students. Be sure each group or pair gets a mix of pleasant and unpleasant smells by labeling the sets "A," "B," "C," and so on. Number each set of bags 1 to 4. Aromatic materials that can be used are perfume, lemon extract, mint extract, vanilla extract, rubbing alcohol, vinegar, or any aromatic oils. Prepare each bag by putting just a few drops of one of the smell-testing samples on two of the cotton balls. Place the cotton in one of the resealable plastic bags and seal the bag.

Presenting the Investigation

1. Introduce the new science terms:
   behavior An observable response in an organism.
   innate behavior An inherited response; not learned.
   learned behavior A response that is acquired by an organism's experience.
   response A reaction of an organism to a stimulus.
   stimulus (plural **stimuli**) Something that causes a response in an organism.

2. Explore the new science terms.
   - *Ethologists* are scientists who study behavior. Ethologists have found evidence that the nervous systems of many animals have parts designed to detect and respond to simple cues or stimuli in their *environment* (everything in the world around you).
   - Innate behavior is *inborn* (controlled by genes received from parents).

- In innate behavior for a certain stimulus, the same response always occurs and occurred the first time the stimulus appeared. For example, eye blinking in response to dry eyes occurs everytime eyes get dry and occurred the first time the eyes were dry.
- Other examples of innate behavior include the changing size of the pupil of the eye as a response to light; reflex action, such the jerking of a hand away from a hot stove; earthworms responding to light by moving to the dark; and the snapping shut of a Venus flytrap in response to touch.
- *Instinct* is innate behavior, which is not limited to a single response but to an orderly sequence of responses, such as a spider spinning a web. There are many orderly steps to this process that the spider did not learn by observation or trial and error. A spider spins a web the first time it tries, thus web spinning is *inborn,* meaning it was passed along from the parents. Also each species spins a particular kind of web.
- Learned behavior is a response that has changed because of experience. A dog sits in response to a command. The dog has learned to do this after many commands and many mistakes.
- Humans can detect a wide range of olfactory (smell) stimuli that is innate.
- Turning one's head away from an irritating odor is instinctive, and *salivating* (secreting saliva inside the mouth) after smelling a favorite food is learned. But each of these behaviors depends on the innate ability to smell as well as other physical abilities for motion.

Did You Know?

Some ingredients of perfume have an offensive odor in themselves. For example, civet, a substance from the anal scent glands of a catlike mammal, smells vile on its own, but is a vital ingredient in many expensive perfumes.

EXTENSION

Ask students to describe what happens when they think about eating a sour pickle or lemon. If they have ever eaten a sour pickle or lemon, students should notice that saliva starts to develop in their mouths just thinking about it. This is a learned behavior, but the ability to produce saliva is innate. If you ask them to think about eating food they have never heard of, there will be no response.

33

Sniffer

PURPOSE

To determine your behavior in reaction to smells.

Materials

4 smell-testing bags
eyedropper
8 cotton balls

Procedure

1. Open bag 1 and hold it in front of your face but not directly under your nose.
2. With your free hand, fan the air above the bag toward your nose.
3. Ask a helper to describe your response to the smell, such as turning your head away from an unpleasant smell or smiling if the smell is pleasant. Fill in your behavior in the Response Data table.
4. Repeat steps 1 to 3 for the other three bags.

Results

Response to smells varies with each individual, but all people tend to respond negatively to unpleasant smells that irritate the nose.

Why?

A **stimulus** is something that causes an organism to react, and a **response** is anything that an organism does in reaction to a stimulus. In this investigation, the stimuli are the different aromatic materials. Special cells in your nose respond to the smell in each bag by sending messages to your brain. This is a response, but it is not a behavior. A **behavior** is an observable response. Facial expressions, such as wrinkling your nose or opening your eyes wider, or other body movements, such as turning your head away in reaction to a smell, are examples of behavior.

Reacting to smells is a combination of **innate behavior** (inherited responses) and **learned behavior** (responses acquired by experience). If the smell is irritating to the nose, the innate response in anyone is to move away from the offending stimulus. A positive or negative response to smells can also be a learned behavior. The smell of foods common to a culture may be offensive to those not used to the smell. For example, the aroma from a pot of fermenting whale blubber may be appetizing to Iñuipiats, but it might not cause an Aborigine's mouth to water!

RESPONSE DATA	
Sniffing Bag	**Behavior**
1.	
2.	
3.	
4.	

Janice VanCleave's Teaching the Fun of Science

Staying Cool

Benchmarks

By the end of grade 5, students should know that

- Organisms make adjustments to heat and cold in order to survive.

By the end of grade 8, students should know that

- Some animals are limited to behaviors determined by inheritance, while others have more complex brains and can learn a wide variety of behaviors.

In this investigation, students are expected to

- Identify a behavior that allows cold-blooded animals to survive.
- Identify the difference between cold-blooded and warm-blooded animals.
- Identify a response in organisms to an external stimulus, such as heat.
- Describe how organisms maintain stable internal conditions while living in changing external environments.

Preparing for the Investigation

Prepare the index cards in advance by folding them in half lengthwise. Then make two slits in each card as shown in the figure in the investigation. The slits need to be large enough for the thermometer to slide through. Student thermometers recessed in plastic to help prevent breakage are recommended. (See appendix 3 for a list of science supply companies where these thermometers can be purchased.)

Presenting the Investigation

1. Introduce the new science terms:

   cold-blooded animal An animal whose internal body temperature changes with the temperature outside its body.

2. Explore the new science terms:
   - *Ectotherm* is another term for cold-blooded animal.
   - Examples of cold-blooded animals are amphibians, including frogs, toads, and salamanders; reptiles, including snakes, lizards, turtles, and crocodiles; and fish.

- Temperature is one of the most important environmental factors affecting ectothermic organisms. These organisms depend primarily on an external heat source and control their temperature by various innate behaviors, such as basking directly in the sun if they are cool or moving until they are out of the sunlight if they are hot. Learned behavior would be if the organism seeks out hot rocks to lean against or areas that are sunny when they are cold or shady, cooler areas or objects when they are hot.

- An animal that generates heat to maintain a constant internal body temperature is called *warm-blooded.*

- *Endotherm* is another term for warm-blooded animals.

- Examples of warm-blooded animals are mammals, including humans, dogs, cats, polar bears, and whales; and birds, including penguins, cardinals, hawks, eagles, and vultures.

Did You Know?

In the desert, a camel sweats so much in order to cool itself that when given a chance to drink, it may consume as much as 50 gallons (187 L) of water in one long drinking binge. This water is not stored in the camel's humps; fat is stored there.

EXTENSION

Humans sweat and dogs pant in order to cool their bodies. This innate behavior occurs in response to an environmental factor—temperature. Sweating or panting cools the body because, in order for water to evaporate from the skin, the water must take in energy. Some of this energy is in the form of heat that the water takes in from the skin. To investigate the cooling effect that evaporation has on the skin, students can rub water on the back of their hands and blow their breath on the wet spot. Point out that their breath is at body temperature so it is not cooling, but it does aid in the evaporation of the water and also pushes away the warmed water vapor above the skin.

Staying Cool

PURPOSE

To determine how the behavior of a lizard can change its body temperature.

Materials

unruled index card with 2 slits
pencil
thermometer
timer

Procedure

1. Fold the index card in half lengthwise.
2. Draw a lizard on the unslit side of the folded card.
3. Insert the thermometer through the slits in the card as shown.
4. Stand the lizard card outdoors so that the thermometer is in direct sunlight.
5. After 3 minutes, read and record the temperature on the card in row 1 of the Temperature Data table.
6. Stand the lizard card in a shady outdoor area for 3 minutes. Again, read and record the temperature.

TEMPERATURE DATA	
Location of Lizard	Lizard's Skin Temperature
in the Sun	
in the shade	

Results

The temperature is lower in the shade than in the Sun.

Why?

Cold-blooded animals are animals whose internal body temperature changes with the temperature outside their body. The behavior of lizards and other cold-blooded animals affects their body temperature. Moving into and out of sunlight will heat and cool their bodies respectively. The higher temperature reading when the thermometer was placed in sunlight indicates that the lizard's skin would have received more heat when the animal stood in a sunny area. The blood in the vessels beneath the skin would have warmed and circulated throughout its body, raising the body temperature of the animal. The lizard moves in response to the sunlight (stimulus) when it gets hot and stops when the stimulus is removed (a shady area). This is innate behavior and is an inborn survival response. Organisms that are capable of learning would learn to distinguish sunlight from shady areas and would seek shady areas when they are hot.

Oops!

Benchmarks

By the end of grade 5, students should know that

- The brain gets signals from all parts of the body telling the brain what is going on in those places.
- The brain also sends signals to parts of the body to influence what those parts do.

By the end of grade 8, students should know that

- The level of skill a person can reach in any particular activity depends on inborn abilities and practice.

In this investigation, students are expected to

- Identify a response in organisms to an external stimulus.
- Understand the process involved in reaction time.
- Identify innate and learned behavior.

Presenting the Investigation

1. Introduce the new science terms:

   cerebellum The part of the brain that controls muscle action.

   cerebrum The largest part of the brain, which controls thoughts.

   hand-eye coordination The ability to move your hand in response to what your eyes see.

   reaction time The time it takes an organism to respond to a stimulus.

2. Explore the new science terms:
   - Reaction time does not depend on how smart you are.
   - The physical ability required for hand-eye coordination is inborn, thus the physical response of catching the cotton ball in the investigation is innate behavior. Since catching cotton balls falling through a tube is not something that is done natu-

rally and requires directions, the response to the falling cotton ball is learned behavior, which can be improved with practice.
   - Doing a task over and over again is a type of learning called trial and error and is the simplest kind of learning. With practice, the act generally improves.
   - Reaction time can improve with practice.
   - In hand-eye coordination, nerves in the eye start the following relay of nerve messages:

     a. The message travels from the eye to the cerebrum.

     b. The cerebrum interprets the message and sends a message to the cerebellum.

     c. The cerebellum interprets the message and sends a message to all the muscles necessary to respond.

     d. The muscles respond.

Did You Know?

Learning occurs when pathways between *neurons* (nerve cells throughout the body) develop. With practice, messages take "shortcuts" through the maze of neurons and new, faster pathways develop.

EXTENSIONS

1. Ask each pair of students to list examples of the ways reaction time can affect the survival of some animals (for example, a deer's response of running when it sees a predator). Discuss the lists and combine the ideas into one list.

2. Ask students for examples of the ways reaction time helps athletes in different sports. For example, in order for baseball players to catch a fly ball, they must run in the right direction, move their hands in the right direction, and close their hands around the ball at just the right time.

Oops!

PURPOSE

To determine whether practice improves reaction time.

Materials

cardboard tube (such as an empty paper towel tube)
ruler
cotton ball
pencil

Procedure

1. Hold one end of the cardboard tube about 4 inches (10 cm) above a table.
2. Hold the cotton ball at the top of the open tube.
3. Drop the cotton ball into the tube.
4. Your helper should watch the bottom of the tube and swat the cotton ball with the ruler when the ball exits the end of the tube.
5. Repeat steps 1 to 4 nine times, recording in the Reaction Time Data table whether the cotton ball is hit during each trial.
6. Change places with your helper and repeat steps 1 to 4.

Results

The number of hits varies with each individual, but usually increases with practice.

Why?

Reaction time is the time it takes an organism to respond to a stimulus. When you see the

cotton ball (the stimulus) coming out of the tube, a message is sent from your eyes to the part of your brain called the **cerebrum** (the largest part of the brain, which controls thoughts. Like a computer, your brain takes this input information and, in fractions of a second, sends a message to the **cerebellum** (the part of the brain that controls muscle action). The cerebellum sends a message telling the muscles in your hand to move. Learning often occurs when multiple stimuli or actions occur at the same time. The more often the combination occurs, the faster the response. So with practice, your reaction time improved and you were able to hit the cotton ball. The ability to move your hand in response to what your eyes see is called **hand-eye coordination.**

REACTION TIME DATA										
	Trials									
Name	1	2	3	4	5	6	7	8	9	10

Ballooning

Benchmarks

By the end of grade 5, students should know that

- Unlike behavior in humans, behavior in some species, such as spiders, is almost entirely genetically determined.

By the end of grade 8, students should know that

- Some animal species are limited to a range of genetically determined behaviors.

In this investigation, students are expected to

- Identify an innate behavior.
- Model the innate behavior of ballooning.

Preparing for the Investigation

You can use other types of labels or stickers, but ¾-inch (1.9-cm) round color-coding labels work well in this investigation. You may wish to use different colors for each group.

Presenting the Investigation

1. Introduce the new science terms:

   ballooning A technique that spiderlings use to move to different places.

   spiderling A young spider.

2. Explore the new science terms:
   - Ballooning is an innate behavior, which means spiders inherit the instinct to do this.
   - Most spiderlings balloon.

Did You Know?

- Spiders have fine hairs on their legs with which they can detect air currents.
- When a spider climbs up its drag line, it balls up the silk thread with its feet and generally eats the ball of silk when it reaches the surface. The silk is digested.

EXTENSION

Spiders have a safety line called a *drag line*. They make this line by attaching a strand of silk to materials they walk on. The making of a drag line is an example of innate behavior. Students can demonstrate a spider's drag line by making spider models out of paper or other material and attaching one end of a thread to the spider and the other end of the thread to a table. The spider can be knocked off the table to demonstrate that the thread holds the spider.

36

Balloforming

PURPOSE

To model the ballooning of spiderlings.

Materials

scissors
ruler
sewing thread
six ¾-inch (1.9-cm) round color-coding labels,
   any color

Procedure

1. Cut 3 pieces of thread, each about 6 inches (15 cm) long.
2. Attach one end of each thread to the sticky side of one of the labels.
3. Fold the label, sticky sides together.
4. Repeat steps 1 to 3 with the remaining labels. These are your six spiderlings.
5. Lay the spiderlings together on a table.
6. Lean toward the table so that your mouth is close to but not touching the spiderlings. Then blow as hard as you can. Observe the movement of the spiderlings.

Results

The spiderlings are moved by your breath to a new area. Some move farther than others.

Why?

Spiders lay their eggs in egg sacs. Some **spiderlings** (young spiders) move away from each other by climbing onto branches and other outdoor surfaces, where they release strands of silk from their bodies. These strands and the attached spiderlings are lifted by wind and float to a new area. This behavior, called **ballooning,** is an innate behavior of most spiders. The labels and thread in this activity represent spiderlings and attached silk strands. As with real spiderlings, the spiderling models are easily moved by wind.

Bent

Benchmarks

By the end of grade 5, students should know that

- Plants can grow more in one direction than the other.

By the end of grade 8, students should know that

- Plant hormones are chemicals that control the growth of plants.

In this investigation, students are expected to

- Identify the plant behavior tropism.

Preparing for the Investigation

Any two colors of paper will work.

Presenting the Investigation

1. Introduce the new science terms:

   auxin A plant hormone that causes changes in the growth of cells.

   plant hormones Chemicals in plants that control cell growth.

   tropism The bending movement of a plant in response to a stimulus, such as light, heat, water, or gravity.

2. Explore the new science terms:

   - All plant behavior is innate.
   - Most plants are stationary, but they are not motionless.
   - Tropism is considered behavior.
   - Tropism occurs because one part of a plant grows faster than another part.
   - The presence of auxin causes some cells to grow longer, such as those in stems, but inhibits the growth of root cells.

Did You Know?

- Some plants move because of an increase in water pressure inside their cells. This pressure is called *turgor pressure*, and the movement is called a *nastic response*. Morning glories open due to nastic responses.

EXTENSIONS

1. Plant behavior helps plants to survive. Ask students to research the main kinds of tropism, phototropism, geotropism, thigmotropism, and hydrotropism. A table can be prepared showing the types of tropism, the stimulus, and an example of the plant response.

TROPISM		
Type of Tropism	**Stimulus**	**Example of Response**
Phototropism	light	Stems bend toward light, allowing leaves to receive light.
Geotropism	gravity	Roots grow down and stems grow up.
Thigmotropism	touch	Vine plants grow around the surface of things, such as poles, allowing them to receive more light.
Hydrotropism	water	Roots grow toward water.

2. Plant motion due to tropism is so slow that it is not seen. But some plants, such as the Venus flytrap and the mimosa move quickly enough to be seen. Ask students to research the movement of these plants to determine why they move. (The responses of both of these plants is due to changes in water pressure. In the mimosa, parts of the leaves contain sensitive cells that lose water when touched. Turgidity in leaves decreases, and the leaves fold very quickly. A similar change occurs in the Venus flytrap in that sensitive cells lose water when touched, causing the leaf to spring shut.)

Bent

PURPOSE

To determine which way plants bend in response to a stimulus.

Materials

scissors
ruler
two 2-by-14-inch (5-by-35-cm) strips of poster
   board—1 white, 1 green
transparent tape

Procedure

1. Cut ¼ inch (0.63 cm) from one end of the white strip of poster board.

2. Lay the green strip on top of the white strip. Tape the ends of the strips together so that they form a "flat loop" as shown.

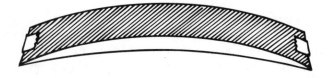

3. Hold the ends of the strips, one in each hand, so that the green strip is on top. Notice how close the surfaces of the strips are.

4. Bend the white strip toward the longer, green side. Observe how close the surfaces of the strips are now.

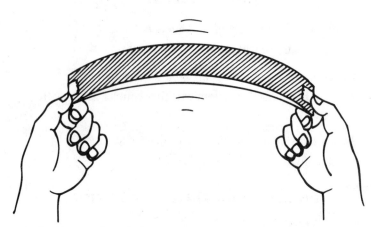

5. Try to bend the green strip toward the shorter, white strip.

Results

The strips bend more easily and are closer when the shorter, white strip bends toward the longer, green strip.

Why?

Plant cells increase in length in response to chemicals called **plant hormones. Auxin** is a plant hormone. In response to different stimuli, such as light, heat, water, or gravity, auxin causes changes in the growth of cells. In plant stems, the presence of auxin causes the cells to grow longer. So when there is a buildup of auxin on one side of the stem, the cells on that side (green strip) grow longer than on the other side (white strip). When this happens, as with the paper strips, the longer side of the stem bends toward the shorter side. A plant's bending movement in response to a stimulus is a plant behavior called **tropism.**

Light Seekers

Benchmarks

By the end of grade 5, students should know that
- Plants grow toward light.

By the end of grade 8, students should know that
- Plant hormones are chemicals that control the growth of plants.

In this investigation, students are expected to
- Identify the plant behavior phototropism.

Preparing for the Investigation

Small potted houseplants will be needed for the experiment. You can provide these or have students bring in their own plants. Using different kinds of plants allows students to compare the behavior of different plants. The experiment could also be done by each student as a home project.

Presenting the Investigation

1. Introduce the new science terms:

    phototropism Growth or movement of a plant in response to light.

    positive phototropism Growth or movement of a plant toward light.

2. Explore the new science terms:
    - Stimuli that affect plant growth include light, heat, gravity, and water.
    - Plants bend toward a stimulus when cells on the side of the stem opposite the stimulus grow longer. This movement is called *positive tropism*.

- Movement away from a stimulus is called *negative tropism*.
- Movement of a plant toward light is called *positive phototropism*.

Did You Know?

- *Thigmotropism* is growth or movement of a plant stimulated by contact with another object, such as vines and other plants that climb or cling to objects. Most climbing plants have weak stems and branches that cannot support their own weight.

EXTENSION

A class project can be designed using the problem "If a plant is sitting near a window, and most of its leaves are bent toward the window, what could you do to make the leaves stand upright?" First, ask students for hypotheses based on known facts. Have the class select one hypothesis, then design an experiment to test the hypothesis. For example, a hypothesis might be "The leaves on a plant will stand up if the plant is turned 180° from the window. This is based on the fact that the leaves on the plant in my investigation bent toward the lighted window." An investigation to test the hypothesis would be to turn the plant so that the label on the plant container faces the window, thus the container is turned 180°. In this new position, repeat the original investigation.

Light Seekers

PURPOSE
To determine how plants respond to light.

Materials
masking tape
pencil
small potted houseplant

Procedure
1. Use the masking tape and the pencil to label one side of the plant container with your name.
2. Set the plant near a window that receives sunlight for at least part of the day. Position the container so that the label faces away from the window.
3. Record the date in the "Starting Day" column of the Plant Data table.
4. Observe the direction of growth of the plant's stems and leaves. Make a drawing of the plant in row 1 of the "Starting Day" column of the table. Include the window in the drawing.
5. Observe the plant each day. When most of the leaves on the plant have bent toward the lighted window, record the date in the "Final Day" column of the table.
6. Repeat step 2, making the drawing in the "Final Day" column of the table.
7. Determine the number of days it took for most of the leaves to turn toward the light.

Results
The leaves on the plant move toward the lighted window. The number of days will vary with type of plant and its number of leaves. It took the author's plant 5 days.

Why?
Growth or movement of a plant in response to light is called **phototropism.** The plant's behavior in this investigation is an example of **positive phototropism,** which is growth or movement toward light. The cells on the side of the plant away from the light grew longer, causing the plant to bend toward its sunny, shorter side.

PLANT DATA		
Position of label	**Starting Day, _____**	**Final day, _____**
facing away from window		

Janice VanCleave's Teaching the Fun of Science

Which Way?

Benchmarks

By the end of grade 5, students should know that

- Things on or near Earth are pulled toward it by Earth's gravity

By the end of grade 8, students should know that

- Plant hormones are chemicals that control the growth of plants.

In this investigation, students are expected to

- Identify the plant behavior gravitropism.

Preparing for the Investigation

Soak four pinto beans per student or group in water overnight. Keep the beans in the refrigerator to prevent souring.

Presenting the Investigation

1. Introduce the new science terms:

   geotropism See **gravitropism.**

   germination The process by which a seed begins to grow.

   gravitropism Growth or movement of a plant in response to gravity; also called **geotropism.**

   negative gravitropism Upward growth or movement of a plant, in the direction opposite the force of gravity.

   positive gravitropism Downward growth or movement of a plant, in the direction of the force of gravity.

2. Explore the new science terms:

   - During germination, roots grow downward and stems grow upward.
   - Plant parts that grow upward, in the direction opposite the force of gravity, exhibit negative gravitropism.
   - Plant parts that grow downward, in the direction of the force of gravity, exhibit positive gravitropism.
   - Plant cells respond differently to auxin. This hormone makes stem cells grow faster and root cells grow slower.

Did You Know?

The banyan tree, which grows in India, is large enough for 20,000 or more people to stand under its branches. Roots of the banyan tree grow down from branches into the soil, forming "pillars" that support the tree. Because of their supporting roots, these are the largest trees in the world.

EXTENSIONS

1. Use the model in the "Bent" investigation 37 to demonstrate how a stem and roots bend toward the side with smaller cells due to unequal growth. Model the bending of stem cells upward by holding the model so that the shorter, white strip is on top. Model the bending of root cells downward by holding the model so that the shorter strip is on the bottom.

2. A small potted houseplant can be used to show gravitropism. Lay the pot on its side in a tray or shallow baking pan. Use pieces of clay to keep the pot from rolling. You can turn the pot upright to water it, but put it back in place as soon as possible. Observe the position of the stems for one or more weeks. The stems will grow upward. When this is observed, stand the pot upright and fill it with water to thoroughly wet the soil in the pot. Allow the plant to sit for about 5 minutes, then carefully remove the plant from the soil. Rinse the roots in a bowl of water. Observe the direction of growth of the roots as compared to that of the stems.

Which Way?

PURPOSE

To determine whether the direction in which seeds are planted affects how they respond to gravity.

Materials

3 paper towels
10-ounce (300-ml) transparent plastic glass
masking tape
pencil
4 pinto beans that have been soaked overnight
tap water
sheet of construction paper

Procedure

1. Fold a paper towel in half and line the inside of the glass with it.

2. Crumple the other two paper towels and stuff them into the glass to hold the paper lining snugly against the sides of the glass.

3. Place a strip of masking tape around the outside of the glass and use the pencil to mark the tape with arrows pointing up, down, left, and right.

4. Place a pinto bean between the glass and the folded paper towel under each arrow. Make sure the curved side of the bean is pointing in the direction indicated by the arrow.

5. Moisten the paper towels with water and keep them moist, but not dripping wet, for the duration of the experiment.

6. Cover the sides of the glass with the sheet of construction paper by wrapping the paper around the glass and taping the edges together. Fold down the top of the paper and secure the edges with tape. This paper cover prevents light from entering the glass.

7. Lift the paper cover and observe the beans daily or as often as possible for 7 or more days.

8. For each day's observation, make a drawing of each bean in the Bean Growth Data table.

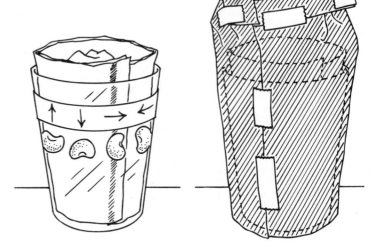

Results

Regardless of how the seeds were planted, the roots grew downward and the stems grew upward.

Why?

Germination is the process by which a seed begins to grow. During germination, gravity pulls auxin down so that it is concentrated in the lower part of the plant embryo. Some cells in the embryo, such as those that develop into the stem, grow faster with the increased presence of auxin. Other cells, such as those that develop into the roots, grow slower with the increased presence of auxin. Growth or movement of a plant in response to gravity is called **gravitropism** or **geotropism.**

Upward growth or movement of a plant, in the direction opposite the force of gravity, is called **negative gravitropism.** Growth of stems is an example of negative gravitropism. **Positive gravitropism** is downward growth or movement of a plant, in the direction of the force of gravity. Roots show positive gravitropism.

BEAN GROWTH DATA

Date	Right ⟶	Left ⟵	Up ↑	Down ↓

Ecosystems and Populations

Ecology is the study of living things and how they interact with each other and their *environment,* which is everything that surrounds them. Plants and animals live together in groups called *communities.* Communities can range in size, from a group of organisms in a rotting tree in a forest to all the organisms in the forest. An *ecosystem* is a community of organisms interacting with their environment or *habitat,* which includes other *biotic* (living) organisms and *abiotic* (nonliving) materials, such as rocks, soil, water, sunlight, and air. Examples of ecosystems are ponds, seashores, meadows, and fields.

The largest ecosystems into which Earth can be divided are called *biomes.* Biomes are large geographical regions identified mainly by the dominant plants that live there. Biomes include tundra, coniferous forests, deciduous forests, grasslands, deserts, rain forests, the ocean, and freshwaters. Animals as well as plants can be used to identify a biome. For example, whales are found only in the ocean. Other identifying characteristics of biomes are their average temperature and moisture. The different biomes students will investigate in this section include tundra, forests, and deserts.

With each community in the natural world, there is a fragile balance in plant and animal *populations* (all the organisms that occur in a specific habitat or that are the same kind or species). In this section, students will examine several ways in which populations' numbers are naturally kept in check.

Treeless

Benchmarks

By the end of grade 5, students should know that

• Plants and animals have features that help them live in different environments.

By the end of grade 8, students should know that

• In any given environment, the growth and survival of plants depend on the physical conditions of the environment.

In this investigation, students are expected to

• Identify one characteristic of a tundra biome to which organisms respond.

Preparing for the Investigation

Use pieces of fettuccine or sticks about 9 inches (22.5 cm) long. The pasta is more easily snapped in half than are sticks, but it may break unexpectedly if not handled carefully. So have extra pieces available. The index card not only represents permafrost, but also prevents the clay from sticking to and staining the table.

Presenting the Investigation

1. Introduce the new science terms:

    active layer The thin layer of ground above permafrost that freezes in winter and thaws in summer.

    Arctic region The region of Earth near the North Pole.

    Arctic tundra The tundra of the Arctic region

    biome A large ecosystem characterized by the plants that occur there due to the climate of the region.

    community Where plants and animals live together.

    ecosystem A community of organisms interacting with their environment.

    environment The conditions around organisms that affect their lives, including weather, land, and food.

    permafrost An underground layer of frozen soil that stays frozen for two or more years.

    tundra A level, treeless biome with year-round air temperature mostly below freezing.

2. Explore the new science terms:

    • The word *tundra* comes from the Finnish word *tunturi*, which means "barren land," but tundras are not generally barren year-round.

• Most tundra plants grow only a few inches (several centimeters) high and are *perennial* (plants whose roots or underground parts continue to grow for 2 or more years).

• Biomes have a special climate because of their location on Earth. In moving from the equator (middle of Earth) to the poles (ends of Earth), the temperature decreases. It is much warmer at the equator than at the poles. The same thing is generally true when going from sea level (low altitude) to the top of a mountain (high altitude).

Did You Know?

• In the Arctic tundra, only about 18 to 24 inches (45 to 60 cm) of the surface of the permafrost thaws each year. Below this, the ground remains frozen.

• Some perennial plants of the Arctic tundra live for decades. It was reported that a rhododendron (an evergreen ornamental shrub) was found whose rings indicated an age of 300 years.

• Buildings in the Arctic are erected on stilts so that the heat from the building does not melt the permafrost beneath the building. If the permafrost melted, the house would sink.

EXTENSION

Tundras are found in high latitudes or at high altitudes all over Earth. Students can research the difference between these types of tundras. *(High-latitude tundras are also called lowland tundras because they are at a low altitude (height above sea level), or arctic tundras because they are found in the high latitudes (imaginary lines dividing Earth into sections from pole to pole) of the Arctic region. The tundra in Antarctica is not well developed and is mainly covered with ice. High-altitude tundras are also called alpine tundras. These tundras do not generally have permafrost.)*

Treeless

PURPOSE

To discover why there are no trees in the tundra.

Materials

index card
grape-size ball of clay
1 piece of fettuccine

Procedure

1. Lay the index card on a table.
2. Divide the clay into two parts, making one part pea-size. Shape each of the clay pieces into a ball.
3. Spread the pea-size ball of clay in the center of the card. Make the clay layer about as big around as a dime.
4. Stick the larger clay ball on the end of the fettuccine. Stand the free end of the fettuccine in the center of the clay on the card, pushing the fettuccine as far as possible into the clay. Be careful not to break the fettuccine.
5. Release the fettuccine and observe any movement.
6. Repeat steps 4 and 5, breaking the fettuccine in half.
7. Repeat step 6 until the fettuccine is short enough to remain standing in the clay.

Results

The piece of fettuccine fell over when made to stand in the clay. A piece of fettuccine that is about 2 inches (5 cm) long or less remains standing.

Why?

A **community** is where people and animals live. An **ecosystem** is a community of organisms interacting with each other and their environment. **Environment** is the conditions around organisms that affect their life, including weather, land, food and nonliving things. A **biome** is a large ecosystem characterized by the plants that occur there due to the climate of the region. The biome called the **tundra** is cold with short plants but is treeless. The tundra in the **Arctic region** (near the North Pole) is called the **Arctic tundra.** One reason the Arctic region is treeless is the presence of **permafrost,** an underground layer of frozen soil that stays frozen for two or more years. The permafrost prevents trees from rooting deep into the soil. Without proper anchorage, any tree that might grow would fall over due to the weight concentrated in the top of the tree or because of wind in the tundra.

In this experiment, the index card represents permafrost and the thin layer of clay represents the **active layer** (a thin layer of ground above permafrost that freezes in winter and thaws in summer). The long piece of fettuccine in the ball of clay represents a tree, and the part of the fettuccine stuck in the clay represents the tree's roots. Since the roots are shallow, they cannot hold up the tree. While tall plants are absent from the tundra, many short plants, represented by the short pieces of fettuccine, grow during the short growing season.

Cones

Benchmarks

By the end of grade 5, students should know that
- Plants and animals have features that help them live in different environments.

By the end of grade 8, students should know that
- In any given environment, the growth and survival of plants depend on the physical conditions of the environment.

In this investigation, students are expected to
- Locate the seeds of pinecones.

Preparing for the Investigation

Small, developed pinecones with tightly closed scales work best. Many of the large, loose-scaled pinecones have lost their seeds. Test the pinecones in advance. If they are too difficult to twist open, soften them by soaking them in water for 2 to 3 hours. Younger students may need assistance even with soaked pinecones.

Presenting the Investigation

1. Introduce the new science terms:

    alpine tundra The tundra of a region above the tree line at high altitudes.

    cone The reproductive structure of a conifer.

    conifer A tree or shrub whose seeds are stored in cones and that usually has needle-shaped leaves.

    coniferous forest A forest made up mainly of conifers and lying below the tree line.

    evergreen Having leaves that are not lost and stay green year-round.

    forest A biome that contains a large group of trees growing close together with various kinds of smaller plants.

    seed cone A cone that contains seeds.

    tree line The border between a forest and a tundra.

2. Explore the new science terms:

    - Coniferous forests lie just south of the Arctic tundra or just below the tree line in an *alpine tundra* (region above about 11,000 ft [3,350 m]).

    - *Boreal forest* and *taiga* are other names for a coniferous forest.

    - Conifers include spruce, pine, fir, cedar, juniper, and redwood.

    - Most conifers are evergreen.

    - Conifers have two kinds of cones: a *pollen cone* that contains male sex cells and forms in groups at the tip of a branch, and (2) a *seed cone* that contains a female sex cell and forms as a single cone or in groups of cones away from the tip of a branch.

    - Generally, spring winds blow pollen grains from the pollen cones to the seed cones. Pollen grains form sperm that fertilize the eggs at the base of the seed cone's scales. Each tree is *bisexual*, meaning that it bears both male and female cones.

    - It can take 2 or more years for some seed cones to develop completely.

Did You Know?

Bristlecone pines are the world's oldest trees. Some living specimens are more than 4,000 years old.

EXTENSION

Tree line or *timberline* is the border between a forest and a tundra. Students can research the location of tree lines. Climate is not the only factor that determines the tree line, but in the Northern Hemisphere it closely follows a line drawn where the summer temperatures average 50°F (10°C). This is near the Arctic Circle, at a latitude of 66½°N. While there is no one specific altitude that marks the tree line for all elevations, it is generally above about 11,000 feet (3,350 m), but can be at lower altitudes.

Cones

PURPOSE

To locate the seeds of a pinecone.

Materials

several sheets of newspaper
1 old washcloth
2 to 3 immature pinecones

Procedure

1. Spread the newspaper on a table.
2. Wrap a washcloth around each end of one of the pinecones.
3. Holding a washcloth-covered end in each hand, twist the pinecone back and forth several times to loosen its scales.
4. While holding the base of the pinecone with the cloth, use the fingers of your other hand to pull out several scales near the tip of the pinecone.
5. Look for two seeds on the inside of each scale, as shown. If you do not find seeds, repeat steps 2 to 4 with another pinecone.

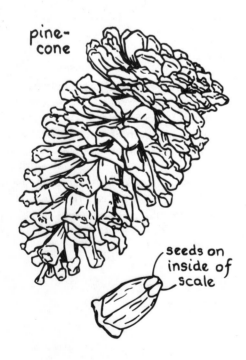

pine-cone

seeds on inside of scale

Results

Two seeds, each attached to a paperlike wing, are found on the inside of each scale of the pinecone.

Why?

A **forest** is a biome that contains a large group of trees growing close together with various kinds of smaller plants. Some forest trees and shrubs, called **conifers,** store their seeds in reproductive structures called **cones,** and usually have needle-shaped leaves. Pine trees are conifers with green, needle-shaped leaves. Pine trees as well as other conifers are found in different areas, but **coniferous forests,** which lie just south of the Arctic tundra or just below the **tree line** (border between a forest and a tundra) in an **alpine tundra,** are made up mainly of conifers. A coniferous forest typically must tolerate dry summers and very cold winters.

Conifers get their name because they make their seeds in cones instead of in flowers. Most conifers are **evergreen,** which means their leaves are not lost and stay green year-round. A pinecone is a **seed cone** (a cone that contains seeds) of a pine tree. Seed cones are able to protect the seeds inside from changes in temperature. When the seeds have completely developed, the scales of pinecones open slightly and the seeds fall to the ground or are blown by the wind.

Seasons

Benchmarks

By the end of grade 5, students should know that

- Some events in nature, such as seasons, occur in a repeating pattern.

By the end of grade 8, students should know that

- The difference in heating of Earth's surface causes seasons.

In this investigation, students are expected to

- Identify the four seasons: spring, summer, autumn, winter.
- Identify the changes in a tree during each of the four seasons.

Preparing for the Investigation

Collect enough empty cardboard toilet tissue tubes so that you have two for each student or group. Make copies of the tree and leaf patterns in advance. Use white, unruled, stiff paper stock to make one copy of the Tree Shapes pattern for each student or group. Use green, red, orange, and yellow copy paper to make a copy of the leaves pattern in each color. If it is not possible to go outside to make a bark rubbing for step 1, use coarse sandpaper. Kids can lay their paper on top of the sandpaper to make the rubbing.

Presenting the Investigation

1. Introduce the new science terms:

    deciduous Having leaves that are lost, usually in autumn.

    deciduous forest See **temperate forest.**

    equator An imaginary line around Earth at 0° latitude that divides Earth into northern and southern halves.

    latitude Distance in degrees north or south of the equator.

    seasons A regularly recurring period of the year characterized by a specific type of weather.

    temperate zone Either of two regions between latitudes 23.5° and 66.5° north and south of the equator.

    temperate forest A forest in a temperate zone; also called **deciduous forest.**

2. Explore the new science terms:

    - Lines of latitude are imaginary parallel lines encircling Earth from the equator to the poles. Latitude is measured in degrees, with the equator being at 0° and the poles at 90°.
    - Oaks and maples are two types of deciduous trees.
    - Temperate forests generally experience four seasons.
    - Trees are divided into three main types: *broad-leaved trees, conifers,* and *palms.*
        - Most broad-leaved trees are deciduous.
        - All broad-leaved trees have flowers, from which they make seeds to grow new trees.

Did You Know?

It is the fewer hours of sunlight, not the cold weather, that trigger deciduous trees to lose their leaves in autumn.

EXTENSIONS

1. Students can research the location of the temperate zones. You may wish for them to locate the temperate zones on a map and/or globe. (There are two temperate zones. One lies between the Tropic of Cancer and the Arctic Circle, and the other between the Tropic of Capricorn and the Antarctic Circle.)

2. Students can also research seasons and learn more about the name of the times each season starts. (Summer and winter solstices and autumn and spring equinoxes.)

Seasons

PURPOSE

To demonstrate changes in a deciduous tree during the four seasons.

Materials

sheet of typing paper
brown crayon
ruler
pencil
scissors
2 cardboard tubes
transparent tape
1 copy of the Tree Shapes pattern
4 copies of the Leaves pattern on colored
    paper—1 green, 1 red, 1 orange, 1 yellow
school glue
2-by-20-inch (5-by-50-cm) strip of pink tissue
    paper

Procedure

1. Lay the sheet of typing paper against the bark of a tree, and rub the brown crayon back and forth across the entire sheet of paper. You have made a bark rubbing.

2. Fold the paper in half, short ends together.

3. Use the ruler and the pencil to mark a 1-inch (2.5-cm) strip along the folded edge of the paper. Cut off this strip and discard it. Two pieces of bark rubbing will be left.

4. Wrap a piece of bark rubbing around each of the tubes, colored side out, and secure with tape. These are the tree trunks.

5. Cut two slits, each about 1 inch (2.5 cm) long, opposite each other on one end of each of the paper tubes (four slits in all).

6. Cut out the spring/summer tree shape.

7. Cut out the green leaves and glue them to both sides of the spring/summer tree shape. (Save 4 to 6 green leaves for step 12, when you make the fall tree.)

8. On one side of the tree shape from step 6, use the tissue paper to add pink flowers. Do this by cutting the pink tissue paper into 1-inch (2.5-cm) squares. Press the eraser of the pencil in the center of one tissue square and squeeze the paper around the pencil. Cover the end of the tissue-covered pencil with glue. Then press the glued end onto the tree shape. Add about 10 of these pink tissue flowers.

9. After the glue has dried, insert each tree shape into the slits in one of the tree trunks.

10. For spring, turn the tree so that the flowers show. For summer, turn the tree so that only the green leaves show.

11. Cut out the fall/winter tree shape. Use the brown crayon to color the branches on one side of the pattern (the winter tree).

12. For fall, cut out the red, yellow, and orange leaves. Glue these to the fall side of the tree shape. Add the remaining green leaves from step 7.

13. Repeat step 9 with the fall and winter trees.

14. For fall, turn the tree so that the colored leaves show. For winter, turn the tree so that the brown branches show.

Results

Models are made of deciduous trees during four seasons.

Why?

Latitude is distance in degrees north or south of the **equator,** which is an imaginary line around Earth at 0° latitude that divides Earth into northern and southern halves. The **temperate zones** are the two regions between latitudes 23.5° and 66.5° north and south of the equator. Forests in the temperate zones are called **temperate forests.** These forests are also called **deciduous forests** because they are

LEAVES

made up of **deciduous** trees, which are trees having leaves that are lost, usually in autumn. The leaves on deciduous trees change color in different **seasons** (regularly recurring periods of the year characterized by specific types of weather). In spring and summer, most deciduous trees have green leaves. Spring flowers are also present on some trees. The leaves change colors, usually to yellow, orange, and/or red in autumn, and fall off, exposing the bare branches in winter.

TREE SHAPES

Spring/Summer

Fall/Winter

Janice VanCleave's Teaching the Fun of Science

Umbrella

Benchmarks

By the end of grade 5, students should know that
- Organisms affect their environment.

By the end of grade 8, students should know that
- Organisms depend on each other.

In this investigation, students are expected to
- Identify how the physical size of tree leaves can affect the environment.
- Identify how one organism affects the survival of others.

Preparing for the Investigation

Collect enough empty cardboard toilet tissue or paper towel tubes so that you have one for each student or group. Make four copies of the Leaf pattern on green copy paper for each student or group.

Presenting the Investigation

1. Introduce the new science terms:

   canopy The umbrellalike top layer of a tropical forest that is formed by the tops of tall broad-leaved evergreen trees.

   humidity Dampness of the air.

   tropical forest A forest in the tropical zone that is made up of tall broad-leaved evergreen trees that form a canopy; also called **rain forest.**

   tropical zone The region of Earth between latitudes 23.5°N and 23.5°S that is characterized by hot, wet weather throughout the year.

2. Explore the new science terms:
   - The average temperature in a tropical forest is 70° to 85°F (21°C to 29°C).
   - The temperature in a tropical forest stays about the same all year, and there is very little change between daytime and nighttime temperatures.
   - *Humidity* is a general term both for dampness of the air and for the amount of moisture in the air, expressed as a percentage.
   - The average humidity in a tropical forest is about 70 percent during the day and 95 percent at night.
   - Trees making up the canopy are generally from 65 to 100 feet (20 to 30 m) tall.
   - The canopy looks like a thick green carpet suspended above the forest.

Did You Know?

Tropical forests cover only about 7 percent of Earth's surface, but about 50 percent of all species are found there.

EXTENSIONS

1. Students can research the location of the tropical zone. You may wish for them to locate the tropical zone on a map and/or globe.

2. Students can research the tropical forest. How much rain does a tropical forest receive? (To qualify as a tropical forest, a forest must receive more than 80 inches [200 cm] per year, but 150 inches [375 cm] is common and some exceed 400 inches [1,000 cm]).

43

Umbrella

PURPOSE

To model the leaves of trees that make up the canopy of a tropical forest.

Materials

scissors
4 copies of the Leaf pattern on green paper
transparent tape
cardboard tube

Procedure

1. Cut out the four copies of the Leaf pattern.

2. Using the tape, secure the stem of each leaf to the top edge of the tube with one stem on each side of the tube. The leaves should stick out from the tube.

3. Close the bottom of the tube by squeezing the sides together and taping them.

4. Stand under an overhead light that is on.

5. Holding the bottom of the tube, lift the tube above your head like an umbrella.

6. Look up at the light and observe how much of the light passes through the umbrella around the edges of the leaves.

Results

A model of a tropical forest tree is made.

Why?

A **tropical forest** is a forest in the **tropical zone,** which is a region at or near Earth's equator, between latitudes 23.5°N and 23.5°S. This forest, also called a **rain forest,** is characterized by hot, humid, wet weather throughout the year. In a tropical forest, the tops of tall broad-leaved evergreen trees form an umbrellalike layer called a **canopy** over the forest. The large and numerous leaves on the trees in this layer act like a huge umbrella blocking most of the sunlight. The forest floor in tropical forests receives very little sunlight. Many of the leaves are pointed, which causes the rain that hits them to run off easily.

The tips are called "drip tips." Because there is so much rain and the canopy acts as a lid over the forest, the air is very damp. This dampness of the air is called **humidity**. Plants that need wet soil and humid air grow well here.

The leaves you made for your tropical forest tree were similar in shape to the leaves of actual tropical forest trees. When you held your tree between yourself and the light, you could see how much light the tree could block.

LEAF

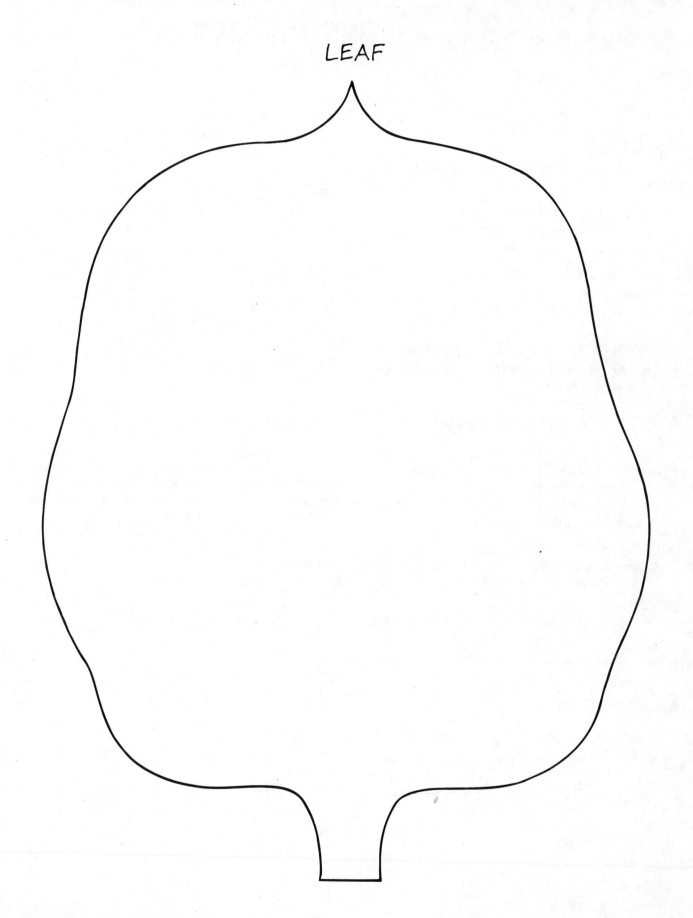

Grass Eaters

Benchmarks

By the end of grade 5, students should know that

- In any environment, some organisms share the same food source.

By the end of grade 8, students should know that

- Animals have eating characteristics that result in their unique niche in an ecosystem.

In this investigation, students are expected to

- Identify a characteristic of a grassland ecosystem to which organisms may respond.

Preparing for the Investigation

- Make one copy of the African Grass Eaters guide for each student or group.

Presenting the Investigation

1. Introduce the new science terms:

   grassland A semiarid biome whose vegetation is mostly grass with few trees or shrubs.

   semiarid Describing a dry climate of low rainfall, but not as dry as a desert.

2. Explore the new science terms:

   - Grasslands get 10 to 20 inches (25 to 50 cm) of rain per year.

   - A semiarid climate is too dry for most trees to grow, but grasses or grasslike plants do well here. Trees and shrubs may be found along streams or in low areas where there is more moisture.

3. A *niche* is the physical location and function of an organism or population within an ecosystem.

Did You Know?

Until the late 1800s, bison (buffaloes) roamed the grasslands of the United States. These animals contributed to the grasslands being free of trees. When they rubbed against the trees on the edge of the grasslands in an effort to shed their winter coat, they stripped the bark from the trees and killed them.

EXTENSION

Students can research the location of the grasslands. You may wish for them to locate the grassland regions on a map and/or globe. They will discover the locations to be in the tropical zone (between latitudes 23.5°N and 23.5°S) and the northern and southern temperate zones (between latitudes 23.5°N and 66.5°N, and 23.5°S and 66.5°S, respectively).

Grass Eaters

PURPOSE

To model the grazing of African animals.

Materials

scissors
1 copy of the African Grass Eaters guide
transparent tape

Procedure

1. Cut around the outside of the African Grass Eaters guide.
2. Fold panel A under along line A. Tape panel A to the back side of panel B.
3. Cut along the dashed lines, cutting through panels A and B.
4. Fold panel C over along line B so that panels B and C face each other.
5. Holding the guide so that the animals are faceup, open each panel to see the part of the grass that each animal eats.

Results

Zebras eat the top, wildebeests the middle, and Thomson's gazelles the bottom part of grass.

Why?

Grassland is a semiarid biome whose vegetation is mostly grass with few trees or shrubs. **Semiarid** describes a dry climate of low rainfall, but not as dry as a desert. Three animals commonly found in Africa's grasslands are Thomson's gazelles, zebras, and wildebeests. The eating habits of these animals are described in this investigation. If a Thomson's gazelle found grass that had not been eaten by other animals, it would not bite the bottom off, leaving the top and middle sections. Neither would a wildebeest eat only from the middle of new grass. But these three animals survive well in grassland because zebras generally graze on the tops of grass and move on to new patches. Wildebeests, eating in an area already grazed on by zebras, graze on the middle of the grass and move on to new patches. This leaves the bottom for Thomson's gazelles to graze on.

AFRICAN GRASS EATERS GUIDE

	Line A ↓		Line B ↓	
Panel A		**Panel B**		**Panel C**

Panel A	Panel B	Panel C
Africa's Grass Eaters	(clouds)	(clouds and sun)
Zebra	**Top**	(grass plant top)
Wildebeast	**Middle**	(grass plant middle)
Thomson's gazelle	**Bottom**	(grass plant bottom)

Stocky

Benchmarks

By the end of grade 5, students should know that

- Living things are found almost everywhere in the world and are somewhat different in different places.
- Differences in their physical structure give organisms an advantage in surviving in a specific habitat.

By the end of grade 8, students should know that

- Individual organisms with certain characteristics are more likely than others to survive.

In this investigation, students are expected to

- Identify a component of a cold desert ecosystem to which an organism may respond.
- Identify a characteristic of penguins that allows them to survive in a cold desert.

Preparing for the Investigation

Any large tray can be used instead of a baking sheet.

Presenting the Investigation

1. Introduce the new science terms:

   cold desert A desert having temperatures that are below freezing for part of the year.

   desert A biome that receives less than 10 inches (25 cm) of rain per year.

   population All the organisms that occur in a specific habitat or that are the same kind or species.

2. Explore the new science terms:
   - The desert biome is one of the most difficult areas for organisms to live in.
   - Antarctica represents one of the harshest biomes: a cold desert.
   - The South Pole receives on average about 2 inches (5 cm) of precipitation per year. Other parts of the continent receive more than this, while some parts have received no measurable precipitation in many years.
   - Parts of Antarctica are the coldest places on Earth, but there are plants and animals living along the coastline, where it is warmer for at least part of the year.
   - Not all penguins live in very cold places, but all live in the Southern Hemisphere. There are 18 different species of penguins.

Did You Know?

- The largest penguin is the emperor penguin, which is about 42 inches (107 cm) tall and weighs more than 90 pounds (41 kg).
- The population of emperor penguins can be in the tens of thousands.
- One way that emperor penguins keep warm is by huddling in a huge group called a *turtle*.

EXTENSIONS

1. You may wish for students to repeat the investigation using a thermometer to measure the temperature difference.
2. All deserts receive a small amount of moisture during the year, but some, such as the Sahara in North Africa and the Mojave in southern California in the western United States, are considered *hot deserts* because of their high temperatures throughout the year. Other deserts, such as the Gobi in northern China or the Atacama along the coasts of Peru and Chile, have freezing temperatures for part of the year. Most of Antarctica has freezing temperatures all year long. Students can research deserts and make a comparison between the types of animals and plants in a hot versus a cold desert. They can identify the different physical characteristics, such as the penguin's stocky body, that help organisms survive in their environment.

Stocky

PURPOSE

To determine why a penguin's stocky shape keeps it warm.

Materials

2 small bowls of warm tap water
baking sheet
timer

Procedure

1. Compare the temperature of the water in both bowls by touching it with your finger.

2. Pour the water from one of the bowls into the baking sheet.

3. Ask a helper to blow on the water in the bowl to cool it off as you blow on the water in the baking sheet. Ask another helper to tell both of you when to start blowing and to stop after 1 minute has passed.

4. Pour the water from the baking sheet back into the empty bowl.

5. Repeat step 1.

Results

The water in the two bowls feels the same at the start of the experiment, but after spreading out one bowl of water on a baking sheet and blowing on each container of water, the water that was spread out feels cooler.

Why?

In order for a material to cool, it must lose heat. The more exposed the surface area of a material, the more quickly it loses heat. If two organisms have the same amount of mass but one has a tall thin body, such as a human, and the other has a short stocky body, such as a penguin, the tall thin human will have more surface area and therefore will lose body heat faster than the short stocky penguin. In this investigation, equal amounts of water were used to represent the same amount of body mass. Like the mass of a tall thin person, the water poured into the baking sheet was spread out and cooled more quickly. Like the mass of a short stocky penguin, the water in the bowl was less spread out and cooled more slowly. So the short stocky shape of penguins' bodies helps them to survive in the cold temperatures of Antarctica.

A large **population** (all the organisms that occur in a specific habitat or that are the same kind or species) of emperor penguins lives along the coast of Antarctica. Antarctica is a type of **desert** (a biome that receives less than 10 inches [25 cm] of rain per year) called a **cold desert,** which has temperatures that are below freezing for part of the year.

Benchmarks

By the end of grade 5, students should know that
- Some animals depend on others for food.

By the end of grade 8, students should know that
- Organisms may interact in a predator/prey relationship.

In this investigation, students are expected to
- Observe and describe the effect of predators and prey in maintaining a balance in population of each.

Preparing for the Investigation

You can prepare the cardboard squares ahead of time. For each student or group, cut five 4-by-4-inch (10-by-10-cm) squares from cardboard boxes.

Presenting the Investigation

1. Introduce the new science terms:

   predator An animal that hunts other animals for food.

   prey An animal that a predator feeds on.

2. Explore the new science terms:
   - A population can be the number of plants in a field or the number of people in a city or in the world.
   - Under normal conditions, the populations of predators and prey are controlled naturally.
   - Sometimes humans disturb this balance by introducing species into the area where they have no natural predators, resulting in the decrease of their prey; or by killing off insects with pesticides, resulting in a decrease in the predators that depend on these insects for survival.

Did You Know?

Predators that hunt in packs have physical gestures to coordinate their activities during a hunt, such as facial expressions or tail movement. Dogs and cats, even though domesticated, still use these inherited behaviors.

EXTENSION

Ask students to research and make a list of predators and their prey. The investigation can be repeated, using cards with different examples of predators and prey drawn on them.

Survival

PURPOSE

To demonstrate one method of natural population control.

Materials

22-by-28-inch (55-by-70-cm) sheet of white poster board
yardstick (meterstick)
20 colored index cards—one color of your choice
five 4-by-4-inch (10-by-10-cm) cardboard squares (cut from cardboard boxes)

Procedure

1. Lay the poster board on the floor. Set a chair 1 yard (1 m) from the edge of the poster board.

2. Randomly spread half of the colored index cards on the poster board. These cards represent snakes, and the poster board the field they live in.

3. Sitting in the chair, toss one of the cardboard squares in an effort to make it land on one of the snakes. The cardboard square represents an eagle.

4. If the eagle lands on a snake, remove the snake from the field and leave the eagle in the field. If the eagle does not land on a snake, remove the eagle and add two more snakes.

5. Repeat steps 3 and 4 until all five eagles have been tossed. Record the number of eagles and snakes in the field in the data column for trial 1 of the Population Control table.

6. Remove and use the eagles from the field to repeat steps 3 to 5, recording the number of eagles and snakes in successive columns of the table.

7. Repeat step 6 until 10 trials have been made or until no eagles remain.

POPULATION CONTROL DATA										
	Trials									
	1	2	3	4	5	6	7	8	9	10
Eagles										
Snakes										

Results

The results will vary, but when there are more eagles, there will be fewer snakes. As the number of snakes decreases, the number of eagles decreases.

Why?

One way that animal populations are controlled is by **predators** (animals that hunt other animals for food). In this activity, the predator is the eagle and the **prey** (an animal that a predator feeds on) is the snake. When an eagle does not find food, it dies, represented by the removal of the eagle from the field. When a snake is not eaten, it can live and reproduce, represented by the addition of two snakes to the field. When there are many snakes in the field, it is generally easier for an eagle to land on one, and the number of snakes decreases. But as the number of snakes decreases, it generally becomes harder for an eagle to land on a snake, so the number of eagles decreases and the number of snakes increases and so on.

Hide-and-Seek

Benchmarks

By the end of grade 5, students should know that

- Differences in their physical structure give organisms an advantage in surviving in a specific habitat.

By the end of grade 8, students should know that

- Individual organisms with certain characteristics are more likely than others to survive.

In this investigation, students are expected to

- Identify characteristics that allow members within a species to survive and reproduce.

Preparing for the Investigation

Students will work independently. In advance, find a large grassy area. Freshly mowed grass is not as good as grass that's a little taller. If you are using the school grounds, notify the school gardener of the area you wish to use and the date so that he or she will not mow the grass before the investigation. Prepare the area and the pipe cleaner "insects" ahead of time. Prepare the area by marking off an area of grass. Place four pencils at the corners of a large square, spacing the pencils about 20 feet (6 m) from each other. Tie one end of a string that is 85 feet (26 m) long to the pencil at the first corner of the square, loop the string around each of the other three pencils, and tie the end of the string to the first pencil. Prepare the pipe cleaner insects using fourteen 12-inch (30-cm) colored pipe cleaners—2 each of black, brown, green, orange, red, white, and yellow. Cut each pipe cleaner into 16 relatively equal-size pieces. On the day of the investigation, without anyone looking, spread around the pipe cleaner insects as evenly as possible in the marked-off area. Keep one insect of each color to show the class. Explain that each color of pipe cleaner represents one kind or species of insect, such as a black beetle, a brown cricket, a green grasshopper, an orange butterfly, and so on. You will be the timekeeper. A watch or stopwatch that marks off minutes will be needed.

Presenting the Investigation

1. Introduce the new science terms:

   camouflage Concealment by protective coloration that helps an animal blend in with its surroundings.

   pattern A consistent arrangement of shapes or colors.

   protective coloration Body coloration or a pattern that helps to camouflage animals from predators.

2. Explore the new science terms:
   - Animals whose colors blend in with their background are said to be camouflaged.
   - Protective coloration helps protect animals from their predators. For example, a bird (predator) that feeds on green grasshoppers (prey) will have trouble spotting the grasshopper on green grass or green leaves. The population of animals with protective coloration is generally greater than that of animals without it.

Did You Know?

- The coloration of most mammals is shades of brown or gray, which is perfect for blending in with their surroundings.
- Most mammals see colors only as variations of light intensity, but mammals with more colorful body coloration generally have at least some color vision.

EXTENSIONS

1. For a math connection, have students determine the percentage of insects found versus that of insects not found. Do this by following these steps:
   - percent of insects found = number of insects found ÷ total insects × 100
   - percent of insects not found = 100% – percent of insects found

2. To make sure all the pipe cleaners are removed from the area so no creature is injured by eating them accidentally, plan a special hunting expedition. Students can imagine that the pipe cleaner pieces are magnetic critters found on another planet and design ways to catch these critters, such as placing a magnet in a sock and dragging it over the area where the critters live.

Hide-and-Seek

PURPOSE

To determine how color camouflage affects insect populations.

Materials

marked-off area of grass with pipe cleaner "insects" spread around in it
paper lunch bag

Procedure

1. Stand outside the marked-off grassy area.
2. When the timekeeper says go, enter the marked-off area and try to find as many of the colored pipe cleaner "insects" as possible. Place the insects in your bag.
3. At the end of 2 minutes, when the timekeeper says stop, immediately stop hunting.

4. Empty the bag of pipe cleaner insects and separate them by color.
5. Count the insects of each color you found and record the number in the Colored Insect Data table.

Results

Depending on the color of the ground in the marked-off area, different amounts of each color will be found. Generally, more of the brighter colors, such as white and yellow, are found.

Why?

A predator is an animal that hunts other animals for food. The animal that becomes the meal for the predator is the prey. Because of special body coloration or **patterns** (consistent arrangements of shapes or colors), some animals blend in with their surroundings and avoid becoming a meal. **Protective coloration** is body coloration or a pattern that helps to conceal animals from predators. This method of concealment by protective coloration is called **camouflage.** In the investigation, you represented a bird eating insects. The color of the piece that you found the least of is the best protective color for insects in the marked-off area. An insect of this color would be less likely to be eaten by a bird than an insect that is the color of the piece you found the most of.

COLORED INSECT DATA							
	Colors						
	Black	Brown	Green	Orange	Red	White	Yellow
Number found							

Too Much, Too Fast

Benchmarks

By the end of grade 5, students should know that

- Solving one problem, such as how to catch more fish, can create other problems, such as extinction due to overfishing.

By the end of grade 8, students should know that

- Technologies often have drawbacks. Some technologies help one group of organisms and harm another.

In this investigation, students are expected to

- Identify the effects of overfishing.

Preparing for the Investigation

Prepare sponge "fish" in advance by cutting 3-by-4-inch (7.5-by-10-cm) dishwashing sponges into as many 1-inch (2.5-cm) cubes as possible. Each student or group needs a bag of at least 40 sponge fish. Make extras in case any group needs them.

Presenting the Investigation

1. Introduce the new science terms:

   extinct No longer in existence.

   overfishing The practice of fishing to such a degree that the fish population is used up.

2. Explore the new science terms:
   - Overfishing by large modern fishing fleets has caused the populations of some fish species to become dangerously low. Atlantic haddock, North Sea herring, and Peruvian anchovy are some of the species that have suffered the effects of overfishing.

Did You Know?

- Since the beginning of time, most extinct animals vanished because they failed to adapt successfully to natural changes in their environment.
- In the last 200 years, and at an increased pace since 1950, some animals and plants have become extinct because of humans. As the human population increases, more food for people and more area for buildings are needed, so plants and animals and their habitats are increasingly being destroyed.

E X T E N S I O N

Ask students to do research to determine the difference between endangered and extinct species. They can make a list of species of each category and determine why each species is endangered or extinct. (An *extinct* species is one that no longer exists. An *endangered* species is one whose population has been reduced to the point that it is at risk of becoming extinct. It could be endangered only in one area; for example, pumas are endangered in parts of the Americas because cattle ranchers kill the pumas, which prey on their cattle.)

Too Much, Too Fast

PURPOSE

To determine the effect that different methods of fishing have on fish populations.

Materials

bag of sponge "fish"
large bowl of tap water
2 tea strainers—1 small, 1 large
small bowl
pencil

Procedure

1. Place 10 sponge "fish" in the large bowl of water. In the Small Net Method Data table, 10 is the population of fish before fishing.

2. Close your eyes and move the small strainer through the water once to scoop up as many sponge fish as possible. Remove the sponge fish from the strainer and place them in the small bowl.

3. Count the sponge fish remaining in the water and record the number in the Small Net Method Data table as the fish population left after fishing.

4. Add an equal number of sponge fish to the water to double the amount of fish in the water.

5. Repeat steps 1 to 4 three times. On the last scooping, count the fish left and double this number to record in the data table, but do not add any fish.

6. Repeat steps 1 to 5 with the large strainer. Record the numbers in the Large Net Method Data table.

Results

The number of sponge fish in the bowl increased when the small strainer was used and fish were added after each scooping. The number of fish greatly decreased and may even have been zero after four scoops with the large strainer.

Why?

The sponge fish represent actual fish, and the strainers commercial fishing nets. Scooping with the small strainer is like fishing with smaller nets and catching fewer fish. Adding sponge fish represents reproduction of fish. Scooping with the large strainer is like fishing with large nets. With a need to provide more fish to feed the growing human population, commercial fishing boats haul in more fish. Large catches are a problem when the fish that are left cannot lay their eggs and reproduce fast enough to keep up the population. **Overfishing** is the practice of fishing to such a degree that the fish population is used up. Some species of fish are in danger of becoming **extinct** (no longer in existence) because of overfishing.

SMALL NET METHOD DATA	
Fish Population Before Fishing	Fish Population After Fishing
1. 10	
2.	
3.	
4.	

LARGE NET METHOD DATA	
Fish Population Before Fishing	Fish Population After Fishing
1. 10	
2.	
3.	
4.	

E

Diversity and Adaptation of Organisms

All living things have some similarities, such as basic cell structure. However, they display *diversity*, or differences. For example, there is a great physical difference between a rose and a penguin, even though both are made of cells that have some similarities. Diversity allows different organisms to live and reproduce in different environments. In this section, students will discover that *adaptation* is an adjustment in a species that enables it to survive in its natural surroundings. Many factors influence how a species becomes adapted to its environment, but survival of the fittest is one of the primary factors. Survival of the fittest does not mean that the toughest or strongest creatures survive; instead it means that organisms that are equipped to handle a particular environmental change will survive better than those that are not. The differences between organisms are due to variations in the traits controlled by genes. If the particular variations needed are there, the individual will be able to survive in the new conditions long enough to reproduce and the offspring will be more likely to have the same variation. For example, moths that have the gene for black coloration survived in areas where buildings were blackened from the pollution due to the burning of coal. The black moths were camouflaged and fewer were eaten by predators. So the population of black moths increased.

Spacey

Benchmarks

By the end of grade 5, students should know that

- Differences in their physical structure give organisms an advantage in surviving in a specific environment.

By the end of grade 8, students should know that

- Individuals with certain traits are more likely to survive and have offspring.

In this investigation, students are expected to

- Analyze how adaptive characteristics help individuals within a species to survive in cold environments.
- Use models to represent an animal's physical adaptive feature and recognize its usefulness in survival.

Preparing for the Investigation

Use stiff copy paper to make one copy of the Caribou Hair pattern for each student or group.

Presenting the Investigation

1. Introduce the new science terms:

    adaptation An adjustment in a physical characteristics or behavior that allows a species to survive in the conditions of its environment.

    insulator A material that is a poor conductor of heat.

2. Explore the new science terms:

    - Coloration and/or pattern is an adaptation that allows some animals to blend in with their background. This type of adaptation is called camouflage. Camouflage helps many organisms survive by allowing them to hide from predators.

- There are millions of different types of adaptations displayed by organisms. A few examples include webbed feet on frogs and ducks, which help them swim; the long claws of the two-toed sloth, which help it grasp tree branches; the shapes of leaves, which help trap sunlight for photosynthesis; and a spider's web, which helps the spider capture insects for food.

Did You Know?

- Polar bears look white, but their hair is actually transparent, hollow tubes. Some sunlight bounces off the hair making it look white. But most of the light passes through the hollow hairs and is absorbed by the black skin to warm the bear.

EXTENSION

Some species live in a *symbiotic relationship* (a close association between organisms of two or more different species that may benefit each member). Ask students to research the relationship between some closely associated species. What is *mutualism*? (A relationship between organisms of two different species in which each member receives some benefit, such as the relationship between the crocodile bird and the Nile crocodile. Another example is the relationship between the cow and the bacteria in its stomach.) What is *parasitism*? (A relationship in which one organism, called a *parasite*, secures its nourishment by living on or inside another organism. The parasite is the *guest* and the organism that it lives on or in is the *host*. The relationship is usually beneficial to the parasite and harmful to the host. Most parasites do not kill their host. Lice and fleas are common parasites that feed off the blood of their host.)

Spacey

PURPOSE

To determine how the adaptive feature of hair shape can help an animal survive in a cold environment.

Materials

scissors
copy of the Caribou Hair pattern
pencil
ruler
10-inch (25-cm) strip of adding machine tape
transparent tape
sheet of construction paper—any color

Procedure

1. Cut out the three caribou hairs from the Caribou Hair pattern.

2. Use the pencil and ruler to draw a line from one end of the strip of adding machine tape to the other. The line should be about ½ inch (1.25 cm) from the edge of the strip.

3. At the left end of the strip, tape the base edge of one of the caribou hairs on the line. Label the "Skin."

4. To the right of the first hair, tape a second caribou hair on the line so that the tips of the hairs meet. Repeat this step with the last hair. A model of a section of caribou skin and hair is made.

5. Lay the model over the colored paper. Observe the amount of colored space between each hair and where the greatest amount of space is.

Results

There is more space between the part of the hairs nearest the skin.

Why?

The reindeer of northern Europe and Asia and the caribou of North America were formerly considered different species. They are now classified as the same species, *Rangifer tarandus*, but the common names are still used to distinguish the two groups. Both of their habitats are very cold, and animals that live there must have some **adaptation** (an adjustment in physical characteristics or behavior that allows a species to survive in the conditions of its environment) to survive the cold. One adaptation is the shape of caribou hair, which is wider at the tip than at the base near the skin. This tapered shape helps to trap air next to the skin of these animals. The colored paper that shows through the model represents spaces between hairs that can be filled with air. Air is a good **insulator,** which is a material that is a poor conductor of heat. So the layer of air helps keep the animals warm. Other animals that live in cold climates, such as the musk ox, have this same hair adaptation.

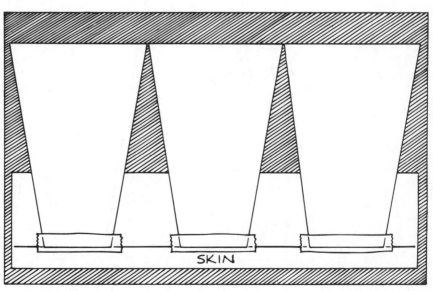

SKIN

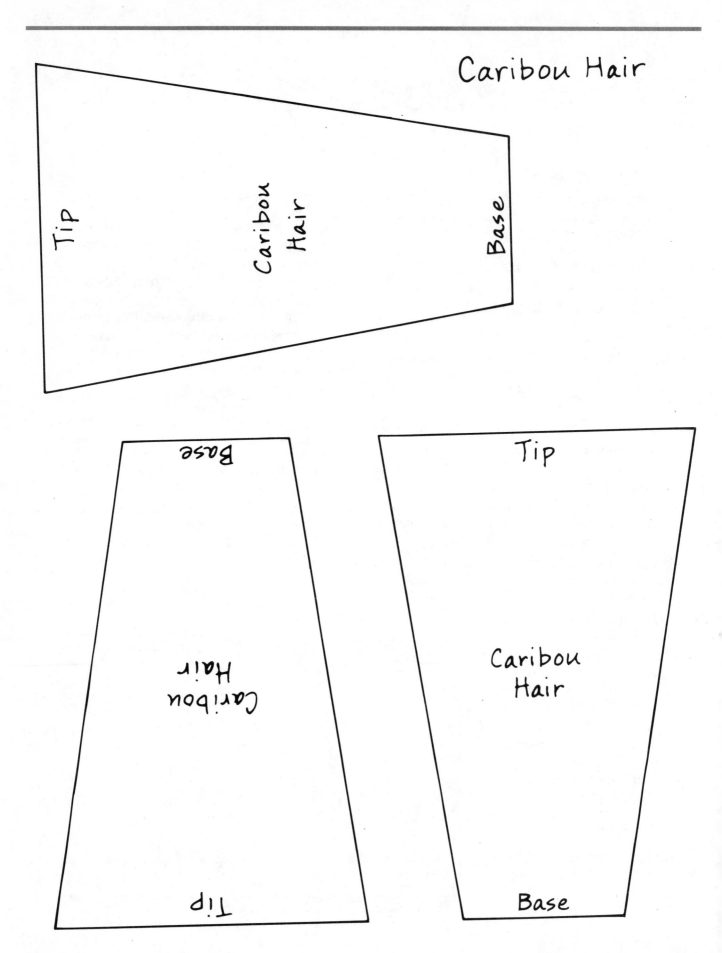

Caribou Hair

Tip

Caribou Hair

Base

Base

Caribou Hair

Tip

Tip

Caribou Hair

Base

Grippers

Benchmarks

By the end of grade 5, students should know that
- Adaptive characteristics of a species improve its ability to survive.

By the end of grade 8, students should know that
- Traits inherited by a species can contribute to the survival of the species.
- Animals have a great variety of body structures that contribute to their being able to find food.

In this investigation students are expected to
- Identify how the adaptive feature of having a thumb affects the behavior of a species.

Preparing for the Investigation

Any large jar that can be picked up with one hand will do.

Presenting the Investigation

1. Introduce the new science terms:

   grip A tight hold or firm grasp on an object.

   mammal A warm-blooded animal that has hair and feeds its young on milk.

   power grip A tight hold on a large object in which the fingers and thumb are wrapped around the object.

   precision grip A tight hold on a small object in which the fingers and thumb pick up the object.

   primate A mammal that has grasping hands with thumbs, including humans, apes, and monkeys.

2. Explore the new science terms:
   - Mammals are *warm-blooded,* which means they generate heat to maintain a constant internal body temperature.
   - Primates have thumbs, which allow them to use a precision grip to pick up small objects and a power grip to hold on to large objects.
   - Like humans, apes and monkeys have hands with five fingers and feet with five toes
   - The feet of apes and monkeys are used much like hands. Using their thumblike big toes, they can grasp with their feet. So their feet are really like an extra pair of hands.
   - Dogs and cats can press things down with their paws, but they cannot use their paws to grip things.
   - Some birds, such as eagles, use the claws on their feet to catch and hold prey, but they cannot lift things to their mouths.

Did You Know?

A parrot, unlike other birds, can grip food with its claws and bring it to its mouth.

EXTENSION

A panda does not have a thumb, but it can tightly grasp things and bring them to its mouth. Ask the class to read about pandas to discover how a panda's paw allows it to grip. (A panda has a special wrist bone called a *radial sesamoid*. When grasping something, such as a bamboo stalk, the panda wraps its five fingers around one side of the stalk, then it pushes the radial sesamoid forward to press the stalk against the fingers.) Students can discover how the radial sesamoid works by repeating the experiment with an object taped across the heel of their hand, such as a marker.

50

Grippers

PURPOSE

To determine the uses of thumbs.

Materials

coin
1-quart (1-liter) jar
adhesive bandage

Procedure

1. Place the coin and jar on a table.

2. Using the fingers and thumb of one hand, try to pick up the coin.

3. Put down the coin and try to pick up the jar by wrapping your hand around it.

4. Hold your thumb against the side of your hand and ask a helper to secure it in this position with the bandage.

5. With your thumb taped down, repeat steps 2 and 3.

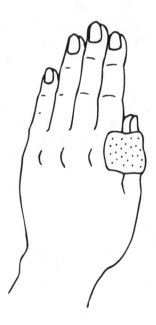

Results

When you were able to use your thumb, it was easy to pick up the coin and the jar. Without the use of your thumb, it was difficult, if not impossible, to pick them up.

Why?

You can pick up small objects, such as coins, because your thumb and fingers press on the object from opposite sides. A tight hold or firm grasp on an object is called a **grip.** A tight hold on an object with your fingers and thumb is called a **precision grip.** When picking up or holding on to large objects, such as the jar, you wrap your hand around the object in what is called a **power grip.** You need the thumb to hold up one side of the jar. **Primates** are **mammals** (animals that have hair and feed their young on milk) that have grasping hands, including humans, monkeys, and apes. Because they have thumbs, they are able to pick up things with a precision grip and to hold objects with a power grip. Not only is having a thumb an adaptation for grasping and for the finer manipulation of objects, but it also permits eating with one hand. In contrast, rodents and other animals must hold their food with two hands.

Sneaker

Benchmarks

By the end of grade 5, students should know that

- Physical features make some animals better predators than others.

By the end of grade 8, students should know that

- Traits inherited by a species can contribute to the survival of the species.

In this investigation, students are expected to

- Identify and explain how a specific characteristic of an organism can affect its predatory skills.

Preparing for the Investigation

If your classroom is carpeted, you may wish to prepare the paper soles in your classroom, then transport the supplies to a tiled area to perform the experiment. A small pillowcase, such as for a baby pillow, works best. If larger pillowcases are used, make sure students secure them properly so they will not trip over loose ends. Rubber bands are added to prevent slipping.

Presenting the Investigation

1. Introduce the new science term:

   absorb To receive sound without echo.

2. Explore the new science term:
   - Soft materials absorb sound better than hard materials. For example, a softer sound is heard when a ball hits a pillow than when it hits a concrete wall.

Did You Know?

Of all the cats in the world, lions are the only ones that live together in large family groups. A family of lions is called a *pride*.

EXTENSION

Point out that lions have other special adaptations that help them to catch their prey. Students can research to discover these features. Some of the features are:

- long, sharp teeth called *canines* for catching and holding prey
- strong claws for grabbing and holding prey
- powerful front leg and chest muscles for knocking down and holding animals as much as three times their size

Sneaker

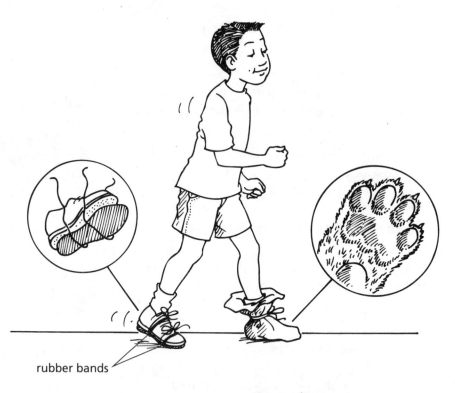

rubber bands

PURPOSE

To determine how the adaptive feature of foot pads makes a lion a more successful hunter.

Materials

pair of hard-soled shoes
12-inch (30-cm) -square piece of
   poster board
pencil
scissors
three 24-inch (60-cm) pieces of
   string
2 rubber bands
small pillowcase

Procedure

1. With the shoes on, place one foot on the poster board. Draw around the sole of the shoe, then cut out the sole shape.

2. With the shoe on, place the paper sole on the bottom of the shoe. Then, secure the paper sole to the bottom of the shoe by tying 2 of the pieces of string around the sole and shoe. *NOTE:* Slip the rubber bands around the paper-covered shoe.

3. Slip your other shoe-covered foot inside the pillowcase and use the remaining string to secure the pillowcase to your shoe. If necessary, fold the ends of the pillowcase over so that they are secured with the string.

4. Take five to six steps across a tiled floor. As you walk, make note of the sound made by your cloth-covered and paper-covered shoes.

Results

   The cloth-covered shoe makes a softer sound than the shoe with a paper sole.

Why?

   Some materials **absorb** (receive without echo) sounds. Soft materials, such as cloth, absorb sounds better than do hard materials, such as poster board. Soft pads on the bottoms of lion's feet, just like your cloth-covered foot, deaden the sound of each step. The foot pads help a lion quietly sneak up on its prey.

Big Ears

Benchmarks

By the end of grade 5, students should know that

- Physical features make some animals better predators than others.

By the end of grade 8, students should know that

- Traits inherited by a species can contribute to the survival of the species.

In this investigation, students are expected to

- Identify and explain how a specific characteristic of an organism can affect its predatory skills.

Presenting the Investigation

1. Introduce the new science terms:

   outer ear The visible outer part of the ear that collects sound waves and directs them inside the ear.

   sound Energy that moves as waves through air or other materials.

2. Explore the new science terms:

   - Large outer ears are an adaptive feature that allows animals to better hear predators, giving them more time to get away.

   - Animals with large outer ears include elephants, deer, zebras, and donkeys.

- Sound energy and radiation both move as waves. Some of the differences between these two types of waves are: (1) Sound waves require a material, while radiation can move through a *vacuum* (empty space). (2) Sound can be heard. (3) Radiation is perceived as warmth or light.

Did You Know?

Deer can turn their ears from side to side in order to catch sounds from all directions.

EXTENSION

You cannot tell the exact direction of the sound source if it is the same distance from both ears. This is because the sound is received with equal intensity by both ears. But if one ear is aimed toward the sound source, more sound waves are received by this ear and the sound is louder. Animals that can twist their ears in different directions can more easily detect the direction of the sound source. Students can design an experiment to determine how moving one's ears helps in detection of the direction of a sound source. One way would be to make two paper cones and hold one against each ear.

Big Ears

PURPOSE

To determine how ear size can help an animal escape a predator.

Materials

sheet of typing paper
transparent tape
ticking clock

Procedure

1. Roll the sheet of paper into a cone shape with a hole about as big as the end of your index finger in the small end and a hole as big as possible in the big end. *Caution:* Do not make the pointed end small enough to stick inside your ear.

2. Secure the cone with tape.

3. Place the clock on a table and stand with one of your ears aimed at the clock. Stand as close as possible to the clock without being able to hear it tick.

4. Hold the small end of the cone up to the ear aimed at the clock. Point the large opening of the cone toward the clock and listen.

5. Remove the cone and listen, then again hold it up to your ear. Compare the difference between the sounds heard with and without the cone.

Results

When you used the paper cone, you were able to hear the ticking of the clock.

Why?

The ticking clock is sending out **sound,** which is energy that moves as waves through air or other materials and can be heard. Your **outer ear** (the visible outer part of the ear) is designed to collect these sound waves and direct them inside your ear. The paper cone enlarged your outer ear, increasing the number of sound waves collected. This made the ticking of the clock loud enough for you to hear it. Large outer ears on animals help them hear predators approaching, which gives them more time to escape.

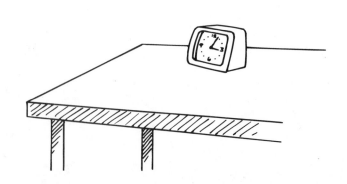

Eagle Eye

Benchmarks

By the end of grade 5, students should know that

- Physical features make some animals better predators than others.

By the end of grade 8, students should know that

- Traits inherited by a species can contribute to the survival of the species.

In this investigation, students are expected to

- Identify and explain how a specific characteristic of an organism can affect its predatory skills.

Preparing for the Investigation

You will need a fairly large open area, as the distance between the student and the ball of clay can be as much as 40 feet (12 m). You may want to make one large data table to record the distances of the entire class.

Presenting the Investigation

1. Introduce the new science term:

   resolving power The eye's ability to focus on objects at a distance.

2. Explore the new science term:
   - The resolving power of an eagle's eye is about eight times that of the human eye.
   - Some eagles may see a rabbit from as far away as 2 miles (3.2 km).

Did You Know?

Eagles have an extra, inner eyelid called a *nictitating membrane* which they pull over their eyes to protect them from being accidentally pecked when their young lunge for food. The membrane also helps to moisten and clean the eyes.

EXTENSION

Point out that eagles have other special adaptations that help them catch their prey. Students can research to discover the following features:

- Some eagles can dive at speeds as great as 200 miles (320 km) per hour.
- Fish eagles, such as bald eagles, have bumps on their toes to help them hold on to slippery fish.

Eagle Eye

PURPOSE

To compare the resolving power of your eyes with that of an eagle's eyes.

Materials

index card
pea-size ball of modeling clay
2 pencils
yardstick (meterstick)

Procedure

1. Lay the index card on a flat piece of ground in a large open area.
2. Place the ball of clay in the center of the index card.
3. Stand in front of the card and look at the clay ball.
4. Back away from the clay ball until you can't see it anymore. Stop at this place and mark it by laying a pencil on the ground.
5. Using the yardstick (meterstick), measure the distance from the pencil to the clay ball. Use the other pencil to record this measurement as d, your resolving power distance, in the Resolving Power Distance Data table. (**Resolving power** is the eye's ability to focus on objects at a distance.)

6. Calculate the resolving power distance of an eagle (d_e) using the following equation:

$$d_e = d \times 8$$

Results

Results vary with each person and the size of the object used.

Why?

The **resolving power** of an eagle's eye is about eight times that of a human eye. If you could see the clay ball from a distance of 38 feet (11.4 m), an eagle could still see it from a distance of 304 feet (91.2 m), a little greater than the length of a football field. An eagle's sharp eyesight is an adaptation that helps the eagle spot prey from on high.

RESOLVING POWER DISTANCE DATA		
Name	d (your resolving power distance)	d_e (eagle's resolving power distance), $d_e = d \times 8$

Gliders

Benchmarks

By the end of grade 5, students should know that

- Animals that migrate have special physical features.

By the end of grade 8, students should know that

- Individuals with certain traits are more likely to survive.
- Migratory animals of different species have different adaptive features.

In this investigation, students are expected to

- Identify and describe how a specific characteristic of an organism can affect its ability to migrate.

Presenting the Investigation

1. Introduce the new science terms:

    glide To move smoothly and effortlessly.

    lift An upward force on a flying object.

    migration Adaptive behavior of moving from one region to another due to changes in environmental conditions.

2. Explore the new science terms:

    - A butterfly is said to glide when it flies without flapping its wings.
    - A butterfly's light body and large wings make it a perfect glider.
    - Butterflies are cold-blooded, which means their body temperature changes with the temperature of their surroundings. When the air temperature gets too cold, some adult butterflies, such as the monarch, migrate to a warmer region.
    - The adaptive behavior of migrating is an *instinct,* which is a complex innate behavior. This behavior is not limited to a single response, but one activity triggers another.
    - Migrating animals do not say to themselves, "It is getting cold and I'd better head south." Instead, they respond to many environmental stimuli, such as the amount of sunlight. This change can affect chemicals in the animal's body, which affects body activity and so on.

Did You Know?

Butterflies can fly if the air temperature is 60° to 108°F (16° to 42°C), but they fly best when their body temperature is 82° to 100°F (28° to 38°C).

EXTENSIONS

1. Students can change the paper gliders to discover how size and shape affect how far the craft will glide and/or how long they will stay in the air.

2. Students can research the migration path of monarchs (from the northern United States to the southern United States and Central America each autumn, returning again in the spring) and the number of miles these small flyers travel (some fly thousands of miles).

Gliders

PURPOSE

To show how monarch butterflies save energy when flying long distances.

Materials

sheet of typing paper
paper clip

Procedure

1. Make a paper glider, following these steps and creasing each fold with your fingernail:
   - Fold the paper in half lengthwise.
   - Open the paper and fold the top corners toward the center so that they touch (figure A).
   - Bring the folded edges toward the center so that they touch (figure B).
   - Again bring the folded edges toward the center (figure C).
   - Turn the paper over and fold along the center so that the outer folded edges (wings) touch. The center fold is the bottom of the body of the glider (figure D).

2. Raise the wings so that they are level, hold the body, and throw the glider to make it fly through the air.

Results

The paper glider flies through the air, then comes to a landing.

Why?

The shape of the paper craft, like the shape of butterfly wings, causes air to flow faster over the top of the wings than below them. The faster-moving air creates an area of lower pressure. The air below the wing has a higher pressure and pushes up on the wing more than the air above the wing pushes downward on the wing. The unequal forces result in an upward force called **lift.** The paper craft slows and gravity pulls it to the surface. However, monarchs are not pulled to the surface because they flap their wings to remain in flight. Adult monarchs are great fliers, flapping slowly but strongly. Between wing flappings, they often hold their wings open and **glide** (move smoothly and effortlessly) through the air. Gliding is a soaring technique that saves energy, allowing monarchs to fly long distances during **migration** (adaptive behavior of moving from one region to another due to changes in environmental conditions).

center fold

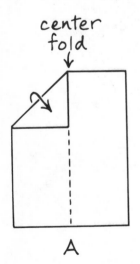
A

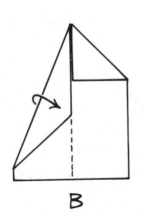
B

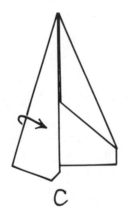
C

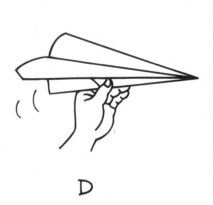
D

Janice VanCleave's Teaching the Fun of Science

Earth and Space Sciences

Earth science is the study of the unique planet we live on: Earth. This science includes information from different sciences, including astronomy, biology, chemistry, physics, and geology. Together these sciences give a better understanding of Earth and its place in space.

Astronomy is space science. This science is the study of *celestial bodies*, which are natural objects in space, including planets, stars, suns, and moons. Astronomy also examines the planet we live on—Earth—and its place among all its celestial neighbors in space.

Structure of the Earth System

The Earth system is made of four interacting parts: (1) the *geosphere* (crust, mantle, and core); (2) the *hydrosphere* (water); (3) the *atmosphere* (air); and (4) the *biosphere* (the realm of all living things). The biosphere was addressed in section III, "Life Science." This section will deal with the geosphere, the hydrosphere, and the atmosphere.

In studying the geosphere, students will learn about the inner layers of Earth, the movement of Earth's solid crust, known as *plate tectonics*, and the erosion processes by which rocks break apart. The different methods that form rocks as well as the changes from one rock form to another, known as the *rock cycle*, will be investigated.

Earth's hydrosphere, or "water sphere," contains all the water on Earth. This part of Earth consists of the world's oceans, lakes, streams, underground water, and all the snow and ice, including glaciers and icebergs. While other planets have a hydrosphere, only Earth has a hydrosphere consisting of water in its three phases—liquid, solid, and gas.

Students will learn about the water of Earth, including the cycle of evaporation and condensation known as the *water cycle*. They will also model a method of measuring the depths of water, known as *sonar*. A model of an iceberg will be made and studied, and the effect of waves on shorelines and the movement of ocean water due to changes in temperature will be investigated.

The blanket of gases surrounding Earth is called the *atmosphere*. The gases in the atmosphere are predominately a mixture of nitrogen, oxygen, carbon dioxide, and water vapor. No other planet in the solar system has an atmosphere comparable to that of Earth's.

Students will learn how the directness of sunlight affects the heating of Earth's atmosphere at the equator. They will demonstrate the effect of Earth's shape on the unequal heating of the atmosphere. The *greenhouse effect* and how materials of Earth's surface affect the greenhouse effect will be investigated. Students will learn about *air* masses and demonstrate how differences in their density cause *warm* and *cold fronts*. How clouds can be used to identify fronts will also be investigated.

Layers

Benchmarks

By the end of grade 5, students should know that

- Scale models show shapes and compare locations of things very different in size.
- Earth is approximately spherical in shape.

By the end of grade 8, students should know that

- The scale chosen for a model makes a big difference in how useful it is.
- Earth is mostly a spherical body made of layers.

In this investigation, students are expected to

- Describe components of the geosphere.

Presenting the Investigation

1. Introduce the new science terms.

   core The inner layer of the geosphere, which is below the mantle and made up mostly of two metallic elements, iron and nickel.

   crust The outer layer of the geosphere, on which organisms live.

   geosphere Solid part of Earth; crust, mantle, core.

   mantle The layer of the geosphere between the core and the crust, which is made up mostly of silicates.

   silicates Chemicals in Earth's mantle that consist of the elements silicon and oxygen combined with another element.

2. Explore the new science terms:
   - The crust is mostly made up of rock.
   - The center of a celestial body is called the *core*.
   - Earth's core, with an average diameter of 4,259 miles (6,800 km), is believed to be made up mostly of two metallic elements, iron and nickel.

- Earth's mantle has a thickness of about 1,812 miles (2,900 km). The most common chemicals found in the mantle are silicates combined with the elements iron and magnesium.
- Earth's crust is 3 to 30 miles (5 to 50 km) in thickness and contains large quantities of silicates combined with aluminum, iron, and magnesium.

Did You Know?

The boundary between each layer of Earth's geosphere is not straight. For example, beneath the Gulf of Alaska there is a rise in the core of at least 6 miles (10 km) into the mantle. And beneath Southeast Asia there is an indentation in the core that is equally deep.

EXTENSION

The temperature and pressure of Earth's layers increase with depth. The high temperatures inside Earth are great enough to melt the materials that make up the layers. Yet most of Earth's interior is not liquid, because the great pressures push the materials together, forming solids and thus offsetting the temperature. Ask students to research the five layers of Earth based on their physical property of phases. These layers and their depths are (1) the *lithosphere* (0 to 100 km), (2) the *asthenosphere* (100 to 350 km), (3) the *mesosphere* (350 to 2,883 km), (4) the *outer core* (2,883 to 5,140 km), and (5) the *inner core* (5,140 to 6,371 km). Repeat the experiment, making a model showing these five layers (lithosphere—solid, asthenosphere—semisolid, mesosphere—solid, outer core—liquid, inner core—solid).

Layers

PURPOSE

To make a model of Earth's three interior layers based on chemical composition.

Materials

drawing compass
metric ruler
6-inch (15-cm) -square piece of cardboard
3 lemon-size pieces of modeling clay—1 each of yellow, red, and blue

Procedure

1. Use the compass and ruler to draw a 6.8-cm-diameter circle in the center of the cardboard.

2. Draw a second circle around the first one with a diameter of 12.6 cm. The circumference of this circle is 2.9 cm from the circumference of the inner circle.

3. Fill the inner circle with yellow clay and the outer ring with red clay.

4. Use the blue clay to make an outline as thin as possible around the ring of red clay.

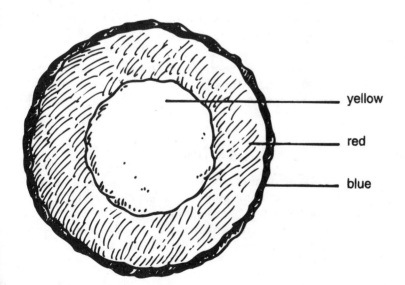

Results

A scale model of Earth's layers is made. The yellow, inner layer of the model is 6.8 cm in diameter; the red, middle layer is 2.9 cm thick; and the blue, outer layer is very thin.

Why?

The three layers of clay represent the three layers of Earth's **geosphere** (solid part of Earth) according to their chemical composition. The inner, yellow layer represents Earth's **core.** Using a scale of 1 cm equals 1,000 km (625 miles), a diameter of 6.8 cm for the yellow layer represents 6,800 km (4,259 miles), which is the average diameter of Earth's core. The core's thickness, or radius, is 3,400 km (2,125 miles). The core is believed to be made up mostly of two metallic elements, iron and nickel.

Surrounding the core is the **mantle.** This layer has a thickness of about 2,900 km (1,812 miles). Thus, the thickness of the red clay representing this layer is 2.9 cm. The most common chemicals found in this layer are **silicates,** which consist of the elements silicon and oxygen combined with another element. The silicates in this layer are mostly combined with the elements iron and magnesium.

The outer layer is called the **crust.** This is the layer you and other organisms live on. This thin outer layer varies from about 5 to 50 km (3 to 30 miles) in thickness. Like the mantle, the crust contains large quantities of silicates, but these are mostly combined with aluminum, iron, and magnesium. On the scale, this means the outer blue clay band representing the crust should be about 0.005 to 0.05 cm (0.002 to 0.02 inches) thick.

Transfer

Benchmarks

By the end of grade 5, students should know that

- Water shapes and reshapes Earth's land surface by eroding rock and soil in some areas and depositing sediments in other areas.

By the end of grade 8, students should know that

- Earth's surface is shaped in part by the movement of water over very long periods of time.

In this investigation, students are expected to

- Identify that the surface of the Earth can be changed by forces such as water and gravity.

Preparing for the Investigation

Make mud balls by mixing soil with water. Shape golf ball–size mud balls and let them air-dry, or bake them in an oven at about 275°F (135°C) for 1 hour or until they are dry. Trays can be cookie sheets, large Styrofoam meat trays, or any containers that will hold the runoff water.

Presenting the Investigation

1. Introduce the new science terms:

    agent of erosion A natural force, such as water, wind, ice, or gravity, that transports eroded materials.

    chemical weathering A type of weathering that changes the chemical properties of crustal materials.

    country rock The common rock of a region.

    erosion The process by which rock and other materials in Earth's crust are broken down and carried away by agents of erosion.

    mechanical weathering A type of weathering that breaks down crustal materials by physical means.

    mineral A solid found in Earth's crust that makes up rocks.

    rock A solid mixture of usually two or more minerals.

    sediments Fragments of a material that have been carried from one place and deposited in another by an agent of erosion.

    weathering The stage of erosion that involves only the breakdown of crustal materials.

2. Explore the new science terms:

    - A mineral has four basic characteristics: (1) it occurs naturally; (2) it is *inorganic*, which means it is not formed from the remains of living things; (3) it has a definite chemical composition; and (4) it has a crystalline structure.

    - Natural agents of erosion include water, wind, ice, and gravity.

    - There are two types of weathering: mechanical and chemical.

    - Common mechanical weathering processes are frost action, wetting and drying, actions of plants and animals, and the loss of overlying rock and soil.

    - One of the main causes of mechanical weathering is the formation of ice in cracks within rock. The ice expands and breaks the rock.

    - Chemical weathering of rock results mainly from the actions of rainwater, oxygen, carbon dioxide, and acids.

    - One of the main causes of chemical weathering is the dissolving action of water.

Did You Know?

The falling water of Niagara Falls is eroding the rock, and the falls are receding. In about 25,000 years, the falls will disappear.

EXTENSION

Freezing water is one way that rocks are weathered. Here's a take-home project that demonstrates this. Have students fill a small plastic container to overflowing with water, secure its plastic lid, and place it in the freezer. When frozen, the lid will bulge with the expanded frozen ice.

Transfer

PURPOSE

To demonstrate water as an agent of erosion.

Materials

masking tape
pencil
three 9-ounce (270-ml) paper cups
tap water
3 mud balls (about the size of golf balls)
   made from soil and water, then dried
tray
ruler

Procedure

1. Use the tape and pencil to label the three cups "A," "B," and "C," then prepare the cups as follows:

   - Use the pencil to make 8 to 10 holes around the bottom edge of the sides of cup A.
   - Use the pencil to make 12 holes in the bottom of cup B.
   - Fill cup C with tap water.

2. Observe the shape of the mud balls and record your observations in the Erosion Data table.

3. Place the mud balls in cup A. Set cup A in the center of the tray. Stand the ruler against the side of cup A and secure it to the cup with tape.

4. Hold cup B 4 inches (10 cm) above cup A. Then pour the water from cup C into cup B as shown.

5. After the water has drained out of cup B, observe the shape of the mud balls in the cup and record your observations in the table. Observe the contents of the tray and record your observations.

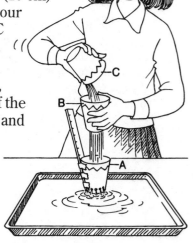

EROSION DATA	
Materials	**Observations**
Shape of original mud balls	
Shade of mud balls after water has been poured over them	
Contents of tray after collecting water from cup	

Results

The water washed away some of the mud from the mud balls and deposited it in the tray.

Why?

Minerals are solids in Earth's crust that make up rocks. **Erosion** is the process by which rock and other materials in Earth's crust are broken down and carried away by natural forces called **agents of erosion.** Before erosion, the common rock of a region is called **country rock.** The stage of erosion that involves only the breakdown of crustal materials is called **weathering.**

In this investigation, the water erodes the mud balls by the two weathering processes. First, the mud balls are **chemically weathered** (a change in the chemical properties of crustal material) by the water dissolving and mixing with the substances in the balls. Second, the balls are **mechanically weathered** (physically broken apart) by the force of the falling water hitting them and breaking away parts. The agents of erosion in this experiment are gravity and water. Gravity pulls the water down, and the water carries the dissolved materials and mixed substances down with it as it flows out of the holes in the cup. When the water stops moving, gravity pulls down the undissolved fragments of the mud balls in the water, and they are deposited on the tray. These fragments that have been carried from one place and deposited in another by an agent of erosion are called **sediments.**

Cemented

Benchmarks

By the end of grade 5, students should know that
- Rock is composed of different combinations of substances.
- Particles of eroded rock are deposited over time, forming layers.

By the end of grade 8, students should know that
- Sediments are gradually buried, then cemented together by dissolved minerals to form solid rock again.

In this investigation, students are expected to
- Understand the process of cementation in the formation of clastic rock, a type of sedimentary rock.

Preparing for the Investigation

Use small natural stones, or aquarium gravel. These can be purchased at garden centers or pet stores.

Presenting the Investigation

1. Introduce the new science terms:

   cementation The binding together of materials.

   clastic rock A type of sedimentary rock formed when sediments from preexisting rock are compacted and cemented together.

   compaction The squeezing together of materials.

   lithification The hardening of sediments into rock.

   sedimentary rock Rock formed of sediments that are deposited by water, wind, or ice.

2. Explore the new science terms:
   - Particle size determines the type of clastic rock.
   - Types of clastic rock are (1) *conglomerate* and *breccia* (made of large rock fragments); (2) *sandstone* (made of grains of sand); and (3) *shale* (made of tiny mineral particles, such as silt and clay).

- Compaction and cementation are stages of lithification in clastic rock formation.
- When country rock is weathered, the sediments are carried by agents of erosion, such as wind and water.
- The sediments deposited in water build up over time and are compacted and cemented together, forming clastic rock.
- The three types of sedimentary rock are grouped according to how they form, and each has different compositions and textures. Compacted and cemented particles form *clastic sedimentary rock*. Deposits of living things form *organic sedimentary rock*. *Chemical sedimentary rock*, called *evaporite*, forms when water evaporates from mineral solutions, such as seawater.

Did You Know?

When sand cements together, it forms a type of sedimentary rock called *sandstone*. Sandstone formations, such as those in Bryce Canyon, Utah, can extend for many miles (kilometers).

EXTENSIONS

1. Conglomerate rock and breccia are made up of various large rock fragments. The fragments in conglomerate are smooth with round edges, while those in breccia are sharp and angular. Students can make samples of conglomerate by repeating the investigation, using large, round, smooth stones.

2. Since deposits of sediments are made at different times, they can vary in content, thus forming sedimentary rock layers. To model sedimentary rock layers, repeat the investigation, adding 2 tablespoons (30 mL) of one color of aquarium gravel and an equal amount of another color of gravel.

Cemented

PURPOSE

To determine how sediments in clastic rock are held together.

Materials

4 tablespoons (120 ml) tap water
two 10-ounce (300-ml) clear plastic glasses
4 tablespoons (60 ml) small stones
school glue

Procedure

1. Pour the water into one of the glasses.
2. Add the small stones to the glass of water.
3. Without pouring out the stones, pour as much of the water as possible out of the glass into the empty glass.
4. Add glue to the glass of stones so that a thin layer of glue covers the top layer of stones. All the stones on the surface should be covered with glue.
5. Squeeze the glass and note how easily the stones move.
6. Allow the glass to sit undisturbed overnight. Then repeat step 5.

Results

Before the glue dried, the small stones could move around, so the cup was easy to squeeze. After the glue dried, the stones were bound together into one solid mass that would not move when you squeezed it.

Why?

In this investigation, the stones represent sediments, and the glue the substances in water that cement sediments together. **Sedimentary rock** is formed of sediments that are deposited by water, wind, or ice. **Lithification** is the hardening of sediments into rock. **Clastic rock** is a type of sedimentary rock formed when sediments from preexisting rock are lithified by the processes of compaction and cementation. During **compaction** (the squeezing together of materials), water is squeezed out of the spaces between the sediments, but substances dissolved in the water may be left behind. During **cementation** (the binding together of materials), these substances form a thin layer around the sediments and bind them together, just as the glue bound the stones in this experiment.

Meltdown

Benchmarks

By the end of grade 5, students should know that
- Substances can change from one phase to another.

By the end of grade 8, students should know that
- One type of sedimentary rock within Earth can be melted and cooled to form another type of rock.

In this investigation, students are expected to
- Determine how rock melts.

Preparing for the Investigation

Chocolate that breaks easily into small squares is best.

Presenting the Investigation

1. Introduce the new science terms:

   igneous rock Rock formed when magma cools and solidifies.

   magma Molten rock beneath Earth's crust.

   melt To change from a solid to a liquid.

   melting point The temperature at which a solid changes to a liquid.

   molten Melted.

2. Explore the new science terms:
   - The temperature of magma is usually 1,022°F to 2,192°F (550°C to 1,200°C)
   - When magma reaches Earth's surface, it is the same but is called *lava*.
   - The size of crystals in igneous rock depends on how fast the molten rock cools. Small crystals are formed when the molten rock cools slowly, and larger crystals when it cools quickly.

Did You Know?

Sometimes solid blocks of different rock above a pool of magma break off and fall into the magma. If the temperature of the magma is below the melting point of these fallen rocks, they remain as solid chunks. These solid pieces of different rock mixed with the molten rock are called *xenoliths*.

EXTENSION

All rocks do not melt at the same temperature. Students can discover this by repeating the experiment, using different kinds of candy, such as caramel, white chocolate, and peppermint. The rate at which each melts can be used to indicate the difference in melting point. A chart can be used to record and compare results.

Meltdown

PURPOSE

How can solid rock melt?

Materials

cup
warm tap water
spoon
timer
½-inch (1.25-cm) -square piece of milk chocolate candy
saucer
toothpick

Procedure

1. Fill the cup with warm tap water.

2. Place the spoon in the cup of water.

3. After about 30 seconds, remove the spoon from the water and place the chocolate in the spoon.

4. Set the spoon on the saucer.

5. Use the toothpick to stir the chocolate in the spoon.

6. Continue to stir the chocolate for 1 minute or until it no longer moves easily.

Results

The chocolate melts, then it becomes hard again.

Why?

Chocolate is solid at room temperature, but like all solids, it **melts** when heated, which means it changes to a liquid. The temperature at which a solid changes to a liquid is called its **melting point.** Because chocolate has a low melting point, the heat of the spoon is enough to raise the chocolate's temperature to the melting point. The hotter the chocolate gets, the more fluid it becomes.

The change in the chocolate from a solid to a liquid due to an increase in its temperature is similar to the change of solid rock to **molten** (melted) rock called **magma** (molten rock beneath Earth's crust). Rocks have a much higher melting point than chocolate. The tremendous heat at depths of about 25 to 37½ miles (40 to 60 km) below Earth's crust is great enough to melt rock. As with chocolate, the hotter the magma, the more fluid it is.

The change of chocolate from liquid to solid is similar to the change of magma to solid rock again. **Igneous rock** is formed when magma cools and solidifies.

Recycled Rock

Benchmarks

By the end of grade 5, students should know that

- There are recognizable patterns of change between different types of rock.

By the end of grade 8, students should know that

- One kind of rock within Earth can be re-formed by pressure and heat to form a different kind of rock.

In this investigation, students are expected to

- Model and summarize metamorphism.

Presenting the Investigation

1. Introduce the new science terms:

    metamorphic rock Rock that forms from other types of rock by pressure and heat within Earth's crust.

    metamorphism The process by which heat and pressure change the makeup, texture, or structure of rocks.

2. Explore the new science terms:

    - Igneous rock and sedimentary rock can be changed to metamorphic rock by heat and pressure. The process by which this occurs is called metamorphism.

    - The rock forming metamorphic rock may be heated until soft and pliable, but it is never melted.

    - Metamorphism of rocks occurs at great depths within Earth's crust or at other areas of high pressure and temperature, such as where plates of Earth's crust are moving. It also occurs near large pools of magma.

    - During metamorphism, the country rock may be heated until it is soft and pliable, and/or pressure causes bits of material making up the rock to be pushed around and compressed. This changes the size and arrangement of the materials in the rock.

Did You Know?

Jade is a translucent ornamental rock used since ancient times in carving. Jade is a product of metamorphism. Jade is found in metamorphic rocks that form when sediments mixed with seawater are pushed beneath the surface by forces deep within Earth.

EXTENSION

A never-ending process by which rocks change from one type to another is called the *rock cycle*. Ask students to research the steps of the rock cycle and make a chart showing the changes from one rock type to another. (Igneous rock forms when sedimentary or metamorphic rock melts, then cools. Sedimentary rock is made from sediments of metamorphic or igneous rock. These sediments form as a result of weathering and are deposited in layers. The layers are compacted and cemented. Metamorphic rock forms when igneous or sedimentary rock is changed by heat and/or pressure.)

Recycled Rock

PURPOSE

To model the formation of metamorphic rock.

Materials

3 slices of bread—2 dark, 1 white
newspaper
two 12-inch (30-cm) -square sheets of waxed
    paper
scissors

Procedure

1. Put the three slices of bread together so that the white slice is sandwiched between the 2 dark slices. Observe the top surface and the edges of the sandwich.

2. Lay the newspaper on the floor.

3. Place one of the sheets of waxed paper on top of the newspaper, and lay the bread sandwich on the waxed paper.

4. Place the other sheet of waxed paper on top of the sandwich.

5. Stand on the paper so that the back of the heel of your shoe is on the sandwich. Shift your weight to this heel, then twist your foot back and forth several times. You want to press the heel of your foot as hard as possible against the bread.

6. Remove the sandwich and cut it in half with the scissors. Observe the top surface and the cut edges of the sandwich.

Results

The three layers of bread are pressed into one thin layer that blends the original layers.

Why?

The original three slices of bread represent three layers of sedimentary rock. When pressure was applied to the model of sedimentary rock by twisting your heel into it, the pressure caused the model to heat up slightly. The pressure and heat on the model represent the forces that are involved when sedimentary rock is changed into another type of rock, called **metamorphic rock** (rock that forms from other types of rock by pressure and heat within Earth's crust). The process by which heat and pressure change the makeup, texture, or structure of rocks is called **metamorphism.**

Crystals

Benchmarks

By the end of grade 5, students should know that

- Rocks are composed of different combinations of minerals.

By the end of grade 8, students should know that

- Rocks bear evidence of the minerals they were formed from.
- Symmetry (or lack of it) may determine properties of crystals.

In this investigation, students are expected to

- Determine the process involved in the formation of the mineral halite.
- Demonstrate that some mixtures maintain the physical properties of their ingredients.
- Identify changes that can occur in the physical properties of the ingredients of solutions, such as the dissolving of salt in water and the evaporation of water from a solution.

Preparing for the Investigation

Each sheet of black construction paper can be cut into four pieces so that each student or group has one piece of paper. Baby food jars or juice glasses are suggested because they are less likely be turned over than are paper cups. Make plans for the placement of the wet experimental papers in advance. Even if some of the water drips when the papers are moved, the results will still be good. The papers will dry with or without being placed in the Sun, but choose a sunny spot if available so that *solar energy* can be discussed.

Presenting the Investigation

1. Introduce the new science terms:

    crystal A solid with flat surfaces that has particles arranged in repeating patterns.

    halite The mineral form of table salt, made of sodium chloride crystals.

2. Explore the new science terms:

    - Examples of crystals are salt, sugar, quartz, and diamond. (If possible, have samples of crystals for students to observe.)

- *Table salt* is the common name for the chemical compound sodium chloride.
- *Halite* and *rock salt* are common names for sodium chloride crystals.
- Minerals can be expressed as a chemical formula. For example, the mineral halite is made of sodium chloride crystals and has the chemical formula NaCl.
- Rocks are usually a mixture of two or more minerals, but halite is a rock made of one mineral.
- A rock can be a mixture of one mineral if it is made up of layers of the same mineral. This happens when seawater evaporates and layer after layer of halite forms.
- Rocks, such as halite, that are formed by the evaporation of water from a mineral solution form *chemical sedimentary rocks* and are called *evaporites*.

Did You Know?

Generally, ice melts as soon as its temperature rises above 32°F (0°C), but a special form of ice crystals called *ice-VII* can be made by putting regular ice under ultra-high pressure so that the atoms and molecules are so close together that it will not melt even at temperatures higher than the boiling point of water, 212°F (100°C).

EXTENSION

Students can determine if the rate of evaporation affects the size of halite crystals. To slow down the evaporation rate, cover the bottom of a bowl with a piece of black construction paper, pour about 1 inch (2.5 cm) of saturated salt solution into the bowl, and allow the bowl to sit undisturbed until the paper is dry. Depending on humidity, this can take 1 or more days. When the paper is dry, students can use a magnifying glass to examine the paper and compare the crystal size to that in the samples from the original investigation.

Crystals

PURPOSE

To discover how salt crystals are formed by evaporation.

Materials

2 tablespoons (60 ml) table salt
4-by-6-inch (10-by-15-cm) piece of black construction paper
magnifying lens
pencil
baby food jar
art brush

Procedure

1. Put a pinch of the table salt on the black paper.
2. Use the magnifying lens to observe the individual salt crystals. Record your observations in the Salt Crystal Data table.
3. Write the first letter of your name on the black paper.
4. Prepare a salt solution by filling the jar half full with water and adding the rest of the salt.
5. Use the brush to stir the salt solution.
6. Paint part of the letter written on the paper with the salt solution. Then stir the salt solution with the brush again and paint another part of the letter. Continue to do this until all the letter is painted.
7. Allow the paper to dry. This may take 30 minutes or longer.

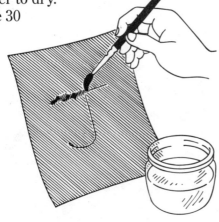

8. After the paper dries, use the magnifying lens to observe the crystals on the paper. Record your observations in the table.

SALT CRYSTAL DATA	
Crystals	**Observations**
Individually	
On the paper	

Results

The individual salt crystals look like small white cubes. The salt crystals on the paper also are cube shaped, but they overlap and may not be as large as the individual crystals.

Why?

Minerals are uniform, crystalline solids found in Earth that make up rock. Table salt is a mineral made of the chemical sodium chloride. The atoms of sodium and chlorine, which make up sodium chloride, are too small to see even with most microscopes. But when many hundreds of thousands of these atoms stack together, they bond, forming a cubed-shaped **crystal** (a solid with flat surfaces that has particles arranged in repeating patterns). The size of salt crystals depends on the number of atoms bonded together. When salt is mixed with water, the atoms of salt break apart. As the water evaporates, there is a smaller amount of water and the separate atoms of salt get closer together. As the water continues to evaporate, eventually the atoms recombine and stack together, forming visible crystals called rock salt or **halite** (the mineral form of table salt, made of sodium chloride crystals). Because the saltwater layer on the paper is thin, the height of the crystals that form is limited.

Spreader

Benchmarks

By the end of grade 5, students should know that
- Some changes in Earth's crust occur over long periods of time.

By the end of grade 8, students should know that
- Earth's crust is made of separate parts.

In this investigation, students are expected to
- Make a model to represent movement of tectonic plates in seafloor spreading.
- Identify forces that shape features of Earth.

Preparing for the Investigation

A square box can be used instead of a round one. Cut the slits in the boxes ahead of time, using a serrated knife. Children should not attempt to cut the slits themselves.

Presenting the Investigation

1. Introduce the new science terms:

    divergent boundary A border where tectonic plates separate and new crustal material is added to the ocean floor.

    lava Molten rock that has reached Earth's surface.

    lithosphere The part of Earth consisting of the crust and the upper part of the mantle.

    midocean ridge One of a number of ridges forming a continuous chain of underwater mountains around Earth.

    rift valley A deep, narrow crack in Earth's crust along the top of a midocean ridge.

    seafloor spreading The process by which new oceanic crust is created and moves slowly away from the midocean ridges.

    tectonic plates Rigid pieces of the lithosphere that cover Earth's surface.

2. Explore the new science terms:
    - Geosphere refers to all of the solid part of Earth and lithosphere to an upper layer. The gaseous and watery part of Earth are the atmosphere and hydrosphere, respectively.
    - The lithosphere extends to an average depth of about 62 miles (101 m).
    - There are seven major tectonic plates. The plate that the United States is on is called the *North American plate*.
    - The tectonic plates rest on a layer in the upper mantle known as the asthenosphere. The *asthenosphere* is made of semisolid material that allows the tectonic plates to more easily shift.

Did You Know?

Although the seafloor is spreading, the total amount of crust stays the same. This is because as new crust is formed at the midocean ridges, old crust sinks into deep-ocean valleys, where it melts and is absorbed back into the mantle.

EXTENSION

The border where two tectonic plates come together is called a *convergent boundary*, and the border where two plates slide in opposite directions beside each other is called a *transform boundary*. Students can use two paper-back books of the same size to model the movement at divergent, convergent, and transform boundaries.

Tectonic Plate Boundaries

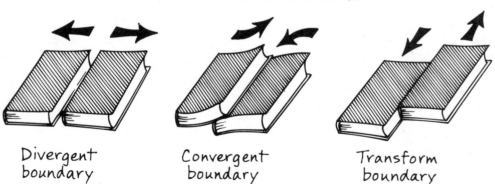

Divergent boundary   Convergent boundary   Transform boundary

Spreader

PURPOSE

To demonstrate seafloor spreading.

Materials

sheet of typing paper
scissors
pencil
ruler
42-ounce (1.19-kg) empty, round oatmeal box
  with a ¼-by-5-inch (0.63-by-12.5-cm) slit in
  the side

Procedure

1. Fold the paper in half with the long edges together.
2. Unfold the paper and cut it in half along the fold line.
3. Using the pencil and ruler, draw a line across each strip of paper about 2 inches (5 cm) from the short end.
4. Put the paper strips together, one on top of the other with the lines together, then insert the unmarked ends of the papers through the slit in the box. When the lines on the strips reach the slit, fold the strips back along the lines to form a pull tab for each strip.
5. Holding one pull tab in each hand, slowly pull about 6 inches (15 cm) of the strips in opposite directions along the surface of the box.

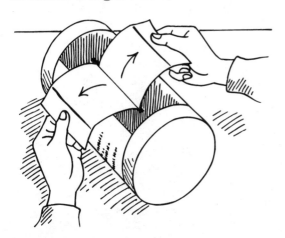

Results

The paper strips emerge from the box and move along the box's surface in opposite directions. The lines get farther apart as the strips are pulled.

Why?

The box represents a **midocean ridge** (one of a number of ridges forming a continuous chain of underwater mountains around Earth). In the center of the midocean ridge is a rift valley. A **rift valley** is a deep, narrow crack in Earth's crust along the top of a midocean ridge, like the slit in the box. Molten rock rises to the surface through this crack. This molten rock, called **lava,** is magma that has reached Earth's surface. About half of the lava rising out of the rift valley spreads on either side of the midocean ridge.

Earth's crust and the upper part of its mantle is called the **lithosphere.** This area is broken into rigid pieces called **tectonic plates,** which cover Earth's surface. The paper strips represent tectonic plates on either side of the midocean ridge. On each strip, the 2-inch (5-cm) section from the end of the strip to the line represents old seafloor bordering the ridge. The remaining part of the strip represents new crustal material that is added to the ocean floor by the lava. The separation of the lines represents a **divergent boundary,** which is the border where the plates separate and new material is added. The lava hardens and forms new ocean floor. The process by which new oceanic crust is created and moves slowly away from the midocean ridges is called **seafloor spreading.** In this investigation, there is a great amount of new material added. But the seafloor may spread only about 1 to 5 inches (2.5 to 12.5 cm) or more per year.

Up and Down

Benchmarks

By the end of grade 5, students should know that
- Some events in nature occur in a repeating pattern.

By the end of grade 8, students should know that
- Water moves in and out of Earth's atmosphere through the water cycle.

In this investigation, students are expected to
- Model the water cycle.

Preparing for the Investigation

Any transparent container will work, such as plastic shoe box–size containers.

Presenting the Investigation

1. Introduce the new science terms:

   condensation The phase change from a gas to a liquid.

   evaporation The phase change from a liquid to a gas.

   precipitation Any form of water that falls from the atmosphere.

water cycle The cycle of evaporation and condensation that moves Earth's water from one place to another.

water vapor Water in the gas phase.

2. Explore the new science terms:

Did You Know?

About 97 percent of all Earth's water is contained in the oceans as salt water.

- Most of the water in the water cycle is provided by the oceans.
- Radiant energy from the Sun drives the water cycle.

EXTENSION

Some of the water in the water cycle comes from plants. The process by which plants give off water by evaporation through porelike openings in the surface of their leaves is called *transpiration*. Students can demonstrate transpiration by placing a clear plastic bag over a group of leaves at the end of a stem of a tree or bush. (Do not cut or break the stem off the plant.) Secure the bag to the stem by wrapping tape around the open end of the bag. Observe the contents of the bag as often as possible for 2 to 3 days. The water leaving the plant's leaves is in the gas phase. Have students describe the series of phase changes in the bag.

Up and Down

PURPOSE

To model the water cycle.

Materials

½ cup (125 ml) warm tap water
2-quart (2-liter) transparent Pyrex bowl
plastic food wrap
ruler
ice cube
resealable plastic sandwich bag
timer

Procedure

1. Pour the warm water into the bowl.
2. Loosely cover the top of the bowl with plastic wrap so that about 2 inches (10 cm) of wrap extends past the edges of the bowl.
3. Put the ice cube in the bag and seal the bag.
4. Place the bag of ice in the center of the plastic wrap that covers the bowl.
5. Gently push the ice down about 1 inch (2.5 cm) so that the plastic wrap sags in the center. Then seal the plastic wrap by pressing its edge against the sides of the bowl.
6. Observe the surface of the plastic wrap directly under the ice cube every 15 minutes for 1 hour or until the ice melts.

Results

At first, the underside of the plastic wrap becomes cloudy and water droplets form under the ice. Over time, the drops under the ice get larger and most of the plastic wrap looks clear. Some of these drops fall back into the water in the bowl.

Why?

Surface water from oceans, lakes, or any body of water is heated by the Sun, and enters the air as a gas called **water vapor.** The phase change from a liquid to a gas is called **evaporation,** and the process requires energy. In the model, the energy needed for evaporation came from the warm water. In nature, the Sun provides the energy.

Water vapor in the bowl rises and comes in contact with the cool surface of the plastic wrap, where it loses energy and forms water droplets. This phase change from a gas to a liquid is called **condensation.** In Earth's atmosphere, water vapor cools and condenses as it rises. The tiny drops of water form clouds that can be moved from one place to the other by winds. Eventually the water in clouds falls as **precipitation** (any form of water that falls from the atmosphere). This cycle of evaporation and condensation that moves Earth's water from one place to another is called the **water cycle.**

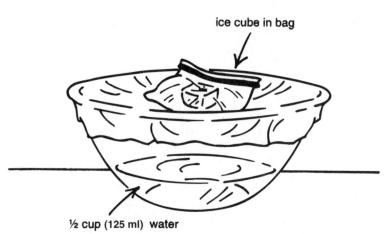

ice cube in bag

½ cup (125 ml) water

Washout

Benchmarks

By the end of grade 5, students should know that
- Water shapes Earth's land surfaces by eroding rock.

By the end of grade 8, students should know that
- Three-fourths of Earth's surface is covered by water, some of which is frozen.
- Earth's surface is shaped in part by the motion of water over very long periods of time.

In this investigation, students are expected to
- Determine how water waves erode a beach.

Preparing for the Investigation

Sand can be purchased where aquarium supplies are sold.

Presenting the Investigation

1. Introduce the new science terms:

   beach A shore with a smooth stretch of sand or pebbles.

   shore The land at a shoreline.

   shoreline A border where a body of water meets the land.

   water wave A disturbance on the surface of water that repeats itself.

2. Explore the new science terms:
   - Beaches vary in width. Narrow beaches may be less than 3 feet (0.9 m) wide, while some wider beaches are over 300 feet (90 m) wide. Not all shores have a beach.
   - Material on a beach varies from sand grains to large rocks.
   - Generally, young beaches are narrow and contain large materials, while older beaches are wider and have smaller materials.

Did You Know?
- Sandy beaches along most of the shores in the United States are made of grains of either granite or sandstone, which are light-colored rocks.
- Black beaches in some parts of Hawaii are formed from the erosion of basalt, a dark volcanic rock.

EXTENSION

A *headland* is a rocky projection from the shore into a body of water. Have students demonstrate the effect of a headland on the erosion of a surrounding sandy beach by repeating the investigation with a rocky headland. Model the headland by placing a large rock on the sand so that it extends into the water. (When there is a headland, fewer waves hit the beach and less sand is washed away.)

Washout

PURPOSE

To simulate erosion of a beach by water waves.

Materials

paint roller pan
4 cups (1,000 ml) sand
2 quarts (2 liters) tap water
pencil

Procedure

1. Cover the bottom of the pan with the sand, making the layer at the shallow end of the pan thicker than the layer at the deep end.
2. Pour the water into the deep end of the pan.
3. Make a mental note of the appearance of the exposed sandy area at the shallow end. This area represents a beach.
4. Make waves by laying the pencil in the deep end of the pan and quickly moving the pencil up and down with your fingertips.

Results

Some of the sand is washed from the beach by the waves.

Why?

The area where the water meets the sand represents a **shoreline** (a border where a body of water meets the land). The **shore** is the land at the shoreline, and a shore with a smooth stretch of sand or pebbles is called a **beach.** As in the model in this investigation, if the beach is mostly sand, it is called a sandy beach.

Moving the water with the pencil causes a disturbance in the surface of the water. A disturbance on the water's surface that repeats itself is called a **water wave.** Water waves that wash against a sandy beach cause erosion of the beach. As in this investigation, ocean waves that wash against a sandy beach move some of the sand into the water.

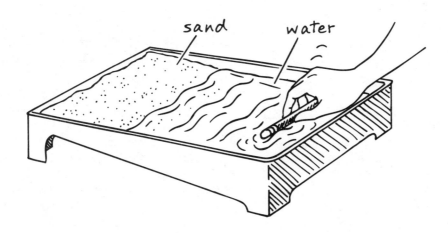

sand   water

Watch Out Below!

Benchmarks

By the end of grade 5, students should know that
- Water can be in the form of a liquid or a solid, and can go back and forth from one form to the other.
- If water is turned into ice, the amount of water is the same as it was before freezing.

By the end of grade 8, students should know that
- Ice takes up more space than does the water from which it is formed.

In this investigation, students are expected to
- Model an iceberg.
- Describe how density affects the position of an iceberg in seawater.

Preparing for the Investigation

The bottoms of plastic soda bottles can be used instead of jars. Cut the tops off the bottles and cover the cut edges with masking tape. You will need access to a freezer to do this investigation.

Presenting the Investigation

1. Introduce the new science terms:

   density The mass per unit volume of a substance.

   glacier A large body of land ice that flows slowly downhill.

   iceberg A piece of a glacier that has broken away and floats in the ocean.

2. Explore the new science terms:
   - Glaciers form when the amount of snow falling in one place is greater than the amount of snow melting. As the snow piles up year after year, the increased weight creates pressure that compresses the snow layers, trapping air in the ice. The surface snow melts and refreezes. The combination of pressure and refreezing turns the compressed snow to ice.
   - The underside of a glacier becomes soft because of pressure, which allows the glacier to move slowly downhill. When the glacier meets the sea, the part that extends into the sea, called a *tongue*, breaks off and forms an iceberg. The process of iceberg formation is called *calving*.
   - Icebergs float because their density is less than that of the water they are in.
   - Ocean water has a great deal of salt and is called *salt water*. Other water sources that have a much lesser amount of salt are called *freshwater*. Salt water has a greater density than freshwater. So there is a greater density difference between ice and salt water than between ice and freshwater. Therefore ice floats higher in salt water than in freshwater.

Did You Know?

The iceberg that sank the *Titanic* in April 1912 was considered a medium-size berg with a length of about 60 feet (18 m) and a height of about 80 feet (24 m) above the water.

EXTENSION

Antarctic icebergs are *tabular* (tabletop-shaped). Have students make a model of a tabular iceberg by shaping a rectangular mold from a 12-by-18-inch (30-by-45-cm) piece of heavy-duty aluminum foil. Fill the foil mold with water and set it on a saucer. Place the saucer in the freezer for 3 hours, or until the water in the box is completely frozen. Fill a 2-quart (2-liter) transparent bowl three-fourths full with water, add 1 tablespoon (15 ml) of table salt, and stir. Remove the ice from the foil mold by peeling away the aluminum. Place the ice in the bowl and observe the amount of ice above and below the surface of the water.

Watch Out Below!

PURPOSE

To demonstrate the position of an iceberg in water.

Materials

3-ounce (90-ml) paper cup
tap water
wide-mouthed 1-quart (1-liter) jar
2 teaspoons (10 ml) table salt
spoon

Procedure

1. Fill the cup with water.
2. Place the cup in the freezer for 2 hours or until the water in the cup is completely frozen.
3. Fill the jar three-fourths full with water.
4. Add the salt to the water in the jar and stir.
5. Remove the ice from he cup. To do this, wrap your hands around the cup for 5 to 6 seconds. The warmth from your hands melts some of the ice, making it easy to remove.
6. Tilt the jar and slowly slide the ice into the jar.
7. Observe the amount of ice above and below the surface of the water.

Results

More ice is below the water's surface than above it.

Why?

When water freezes, its molecules expand, moving farther apart. Ice has the same mass as liquid water, but because it has a greater volume, ice has a slightly lower **density** (mass per unit volume of a substance) than water. As a result, ice floats in water. **Icebergs** are pieces of a **glacier** (a large body of land ice that flows slowly downhill) that has broken away and floats in the ocean. Icebergs, like the ice in this experiment, float in seawater, which is salty. And like all floating ice, most of the ice is below the surface.

Air Blanket

Benchmarks

By the end of grade 5, students should know that
- Air is a substance that surrounds Earth.

By the end of grade 8, students should know that
- Earth is surrounded by a relatively thin blanket of air.

In this investigation, students are expected to
- Draw a diagram of Earth's atmosphere.

Presenting the Investigation

1. Introduce the new science terms:

   atmosphere The blanket of gases surrounding a celestial body.

   exosphere The outermost layer of the atmosphere that extends into space.

   mesosphere The layer of Earth's atmosphere between the stratosphere and the thermosphere.

   stratosphere The layer of Earth's atmosphere between the mesosphere and the troposphere.

   thermosphere The layer of Earth's atmosphere between the mesosphere and the exosphere.

   troposphere The lowest, innermost layer of Earth's atmosphere.

2. Explore the new science terms:
   - The main composition of Earth's atmosphere is nitrogen (78 percent) and oxygen (21 percent). The gases in the remaining 1 percent are argon, carbon dioxide, water vapor, and traces of other gases, including ozone.
   - Below the exosphere, the atmosphere is divided into four regions classified by temperature. The temperature decreases with altitude in the troposphere and mesosphere and increases with altitude in the stratosphere and thermosphere.
   - The troposphere contains over half of all the air in the atmosphere.
   - The temperature in the troposphere decreases with height, which is why the air is cooler on mountaintops.
   - Almost all weather occurs in the troposphere.
   - *Weather* is the state of the atmosphere at a particular time and place. The elements of weather include wind, temperature, precipitation, and atmospheric pressure.
   - The thermosphere is also referred to as the *ionosphere* by some scientists. Ultraviolet radiation from the Sun causes atoms and molecules in this region to lose electrons and become charged particles called *ions*.
   - Above the thermosphere is the *exosphere,* which extends to about 6,000 miles (9,600 km) above Earth. The *magnetosphere* surrounds the exosphere.
   - The altitudes given for the atmospheric layers are averages. The thickness of the atmospheric layers is greatest above Earth's equator and least above the poles.

Did You Know?

The atmosphere acts as an insulator. Without its atmosphere, some places on Earth would be about 176°F (80°C) during the day and –220°F (–140°C) at night.

EXTENSION

The troposphere is the layer that is most familiar because it is the one that directly surrounds us. You may wish to ask students to discover what events happen in the other atmospheric layers. They can add drawings to their atmosphere diagrams to show this. You may wish to ask questions such as these:

1. Where do auroras occur? (ionosphere or thermosphere, because of the charged particles)

2. Where do most meteors burn up? (mesosphere, because of the increase in the density of air molecules)

3. Where is the ozone layer? (upper stratosphere)

4. With an increase in altitude, why does the temperature of the thermosphere and stratosphere increase? (Being the outer layer, more solar ultraviolet radiation (UV) enters this region. There is more UV to absorb in the upper thermosphere. Temperature decreases at lower levels because there is less UV to absorb. But in the stratosphere there is *ozone* [oxygen atom triplet, O_3] which is a better absorber of ultraviolet radiation than ordinary oxygen [two-atom, O_2].)

Air Blanket

PURPOSE

To draw a diagram of Earth's atmosphere.

Materials

pen
ruler
sheet of white copy paper

Procedure

1. Use the pen and ruler to draw a 6-by-9-inch (15-by-22.5-cm) rectangle on the paper.
2. Starting at the top of the rectangle, draw five dashed lines 1½ inches (3.75 cm) apart across the rectangle.
3. Using the pen and ruler, draw a line 1 inch (2.5 cm) from the left side from top to bottom on the rectangle.
4. Write "Atmosphere" above the top line and "Earth" below the bottom dashed line. Add the names of the atmospheric layers in order from bottom to top as shown: Troposphere, Stratosphere, Mesosphere, Thermosphere, Exosphere.
5. Add the altitude measurements shown and an arrowhead at the top of the line.
6. Draw some land features of Earth extending above the 0 mile (0 km) altitude.

Results

You have drawn a diagram of Earth's atmosphere.

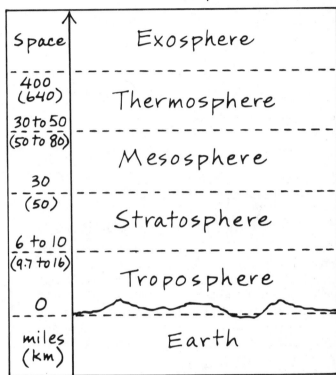

Why?

The Earth's **atmosphere** (the blanket of air surrounding a celestial body) is divided into five basic layers. Starting with the lowest, innermost layer, they are the **troposphere,** the **stratosphere,** the **mesosphere,** the **thermosphere,** and the **exosphere.** There are differences in the layers, including distance from Earth, presence of air, temperature, and what happens there. While the diagram shows a specific starting point for each layer, there are no real barriers dividing the layers and the layers do vary in distance from different points on Earth. For example, at Earth's equator, the troposphere extends to about 12 miles (19 km), while at the poles it extends only about 5 miles (8 km).

Trapped

Benchmarks

By the end of grade 5, students should know that

- Air is a substance that surrounds Earth.
- The Sun is the main source of energy for Earth.

By the end of grade 8, students should know that

- Earth is surrounded by a relatively thin blanket of air.
- Energy from the Sun is available indefinitely.

In this investigation, students are expected to

- Demonstrate the greenhouse effect.
- Determine the causes of the greenhouse effect.

Preparing for the Investigation

Shoe boxes can be placed in a window that receives direct sunlight instead of being placed outdoors. Thermometers need to be student thermometers with safety holders. See appendix 3 for a list of science catalog supplies of these thermometers.

Presenting the Investigation

1. Introduce the new science terms:

    greenhouse effect The warming of Earth by gases in the atmosphere, which trap infrared radiation from the Sun and reradiate it toward Earth, just as the glass or plastic of a greenhouse traps and reradiates infrared radiation within it.

    reradiate To emit previously absorbed radiation.

2. Explore the new science terms:

    - The amount of carbon dioxide in the air from the burning of *fossil fuels* has increased greatly over the last 100 years. This increase could lead to an increase in Earth's average temperature as a result of the greenhouse effect.
    - Fossil fuels (coal, oil, and natural gas) are energy sources made from buried remains of decayed plants and animals that lived hundreds of millions of years ago.

Did You Know?

About 30 percent of the Sun's total radiant energy reaching Earth is reflected back into space by the atmosphere, clouds, and Earth's surface. About 20 percent is absorbed by the atmosphere, and the remaining 50 percent is absorbed by Earth's surface.

EXTENSION

Do surface materials affect the greenhouse effect? Students can find the answer by repeating the experiment. Prepare boxes with different surfaces by covering the soil with different materials, such as sand, rocks, and grass. A surface of water could be prepared by lining the box with plastic and filling it with about 2 inches (5 cm) of water. (Yes, surfaces do affect the greenhouse effect. Light-colored surfaces, such as white sand, reflect from 15 to 45 percent of solar radiation, while darker surfaces, such as grass, reflect 10 to 30 percent. The darker the surface, the less radiation reflected.)

Trapped

PURPOSE

To demonstrate the greenhouse effect.

Materials

2 shoe boxes
ruler
soil
2 thermometers
colorless plastic food wrap
timer
pencil

Procedure

1. Cover the bottom of each shoe box with about 2 inches (5 cm) of soil.

2. Lay a thermometer on the surface of the soil in each box.

3. Cover the opening of one box with a single layer of plastic wrap. Leave the other box uncovered.

4. Take the readings from both thermometers and record them in the Temperature Data table.

5. Place both boxes side by side in a sunny place outdoors.

6. Record readings from both thermometers every 15 minutes for 1 hour in the table.

Results

The temperature readings show that the temperature inside the plastic-covered box was higher and increased faster than the temperature inside the uncovered box.

Why?

Radiant energy from the Sun passes through Earth's atmosphere and reaches Earth's surface. Radiant energy absorbed by Earth is changed into infrared radiation, an invisible form of radiant energy that has a heating effect. Some of this heat from Earth warms the cooler atmosphere above it. Heat is transferred to the atmosphere by three processes: conduction, convection, and radiation. Carbon dioxide and water vapor are gases in the atmosphere that help keep heat from being lost to space. These and other gases absorb heat from Earth and **reradiate** (emit previously absorbed radiation) heat toward Earth. Like the plastic covering that prevents the escape of some of the infrared radiation from the soil, Earth's atmosphere keeps Earth warm. Because the atmosphere helps warm Earth's surface by trapping infrared radiation, the atmosphere is similar to a greenhouse. Hence the term **greenhouse effect.**

TEMPERATURE DATA					
	Time (minutes)				
Container	At start	15	30	45	60
Covered box					
Uncovered box					

Boundaries

Benchmarks

By the end of grade 5, students should know that
- Air is a substance that surrounds Earth, and whose movement is felt as wind.

By the end of grade 8, students should know that
- Earth is surrounded by a relatively thin blanket of air.

In this investigation, students are expected to
- Model a front.
- Identify the boundaries of air masses of different temperatures.

Preparing for the Investigation

You can use a different size plastic bottle. Just adjust the amount of water and oil so that the bottle is about half full of liquid.

Presenting the Investigation

1. Introduce the new science terms:

   air mass A large body of air that is about the same temperature throughout.

   front A boundary between cold and warm air masses.

2. Explore the new science terms:

Did You Know?

Vilhelm Bjerknes (1862–1951), a Norwegian physicist and meteorologist, coined the term *front* to describe the boundary between warm and cold air masses.

- The temperature differences between adjoining regions of Earth result in movement of air masses and changes in local weather.
- Masses of air that stay in place for a length of time take on the temperature of the region beneath them.

EXTENSION

Ask students to research fronts. What is the difference between warm, cold, stationary, and occluded fronts? What effect do fronts have on weather? (The leading edge of a warm air mass advancing into a region occupied by a cold air mass is called a *warm front*. A *cold front* occurs when a cold air mass advances into a region occupied by a warm air mass. If the boundary between cold and warm air masses doesn't move, it is called a *stationary front*. The boundary where a cold front undercuts a warm front, pushing air upward, is called an *occluded front*. At a front, the weather is usually unsettled and stormy, and precipitation is common.)

Boundaries

PURPOSE

To model a front.

Materials

1-cup (250-ml) measuring cup
tap water
blue food coloring
spoon
20-ounce (600-ml) clear plastic bottle
1 cup (250 ml) cooking oil

Procedure

1. Fill the measuring cup with water.
2. Add three drops of food coloring to the water and stir.
3. Pour the water into the bottle.
4. Fill the measuring cup with oil.
5. Tilt the bottle of water and slowly pour the oil into the bottle.
6. Observe the movement of the oil in relation to the water as the oil flows into the bottle.

Results

The oil moves across the top of the blue water.

Why?

An **air mass** is a large body of air that is about the same temperature, pressure, and humidity throughout. Air masses form when air stays over a region long enough to take on the temperature of the region. It takes a week or more for an air mass to form.

Temperature is one of the things that affects the density of an air mass. Warm air masses are less dense than cold air masses. When air masses with different densities meet, the two masses do not mix. Like the oil and the water, a distinct boundary, called a **front,** forms between the air masses. In this investigation, the oil represents a warm air mass and the colored water a cold air mass. Like the oil and the water, warm, less dense air moves over colder, denser air.

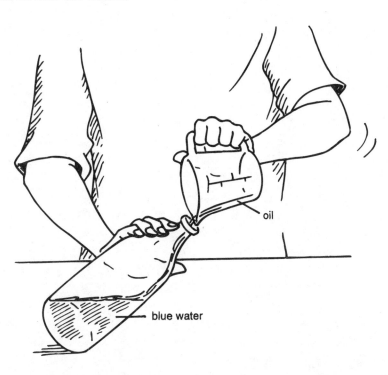

oil

blue water

Earth in Space

We are passengers on a giant spaceship called *Earth*. Our spaceship spins like a top as it rushes through space. But because Earth is so large, and everything on it spins and moves at the same speed, we don't sense this motion.

In this section, students will investigate the motion of Earth and how it affects the apparent motions of celestial bodies, including the Sun, the Moon, and the stars. The effect of Earth's motion on how we reckon days, years, and seasons will be determined. The location and size of star groups will also be studied.

Sun Parts

Benchmarks

By the end of grade 5, students should know that
- The Sun is the main source of heat for Earth.

By the end of grade 8, students should know that
- The Sun has dark spots.
- Heat from the Sun is available indefinitely.

In this investigation, students are expected to
- Model and identify the layers of the Sun.

Preparing for the Investigation

Different colors of modeling clay could be used instead of the Styrofoam ball. For safety, cut the Styrofoam balls in advance, using a serrated knife to remove one-fourth of each ball. The section removed should be wedge shaped.

Presenting the Investigation

1. Introduce the new science terms:

   convection zone The layer of the Sun between the radiation zone and the photosphere.

   core The center of a celestial body. The sun's core is the hottest part of the Sun.

   nuclear fusion Joining of the nuclei of atoms.

   photosphere The outermost layer of the Sun, which is around the convection zone and is actually the first layer of the Sun's atmosphere.

   radiation zone The layer of the Sun between the core and the convection zone.

2. Explore the new science terms:
   - The Sun is made of hot gases, mostly hydrogen with some helium and other elements.
   - The core is the hottest part of the Sun. It is in the core that nuclear fusion occurs.
   - Nuclear fusion produces the Sun's radiant energy, including heat.
   - Heat from the Sun's core is transmitted through the radiation zone.
   - From the radiation zone, gas expands and rises, then cools, becomes denser, and sinks down. This circulating gas forms the convection zone.
   - The layers of the Sun's atmosphere, outward from the Sun's surface, are the *photosphere*, the *chromosphere*, and the *corona*.

Did You Know?
- The Sun is 93 million miles (149 million km) from Earth.
- Light travels at a speed of about 186,000 miles (300,000 km) per second.

  A light-year is a distance unit describing how far light travels at this speed in 1 year. A light-year is about 6 trillion miles (9.5 trillion km). It takes about 8½ minutes for light from the Sun to reach Earth. The next closest star is Alpha Centauri, and its light takes about 4.3 light-years to reach Earth.

EXTENSION

Students can make a legend for the model that gives each layer's thickness and temperature.

THE SUN'S LAYERS		
Layer	**Thickness, miles (km)**	**Temperature, °F (°C)**
Core	87,000 (139,200)	27,000,000 (15,000,000)
Radiation zone	239,250 (382,800)	4,500,000 (2,500,000)
Convection zone	108,750 (174,000)	1,980,000 (1,100,000)
Photosphere	342 (547)	9,932 (5,500)

Sun Parts

PURPOSE

To prepare a model of the Sun's structure.

Materials

2 permanent markers, each a different color
6-inch (15-cm) or larger Styrofoam ball from
   which one-fourth has been removed
four 1-by-4-inch (2.5-by-10-cm) white labels
4 round toothpicks
pen

Procedure

1. With one of the colored markers, paint an area in the center of the cutaway section of the ball to represent the Sun's core as shown.

2. Use the other colored marker to draw a band inside the cutaway section to represent the convection zone as shown.

3. Prepare flags as shown, following these steps:
   - Put just the ends of the sticky sides of a label together. Do not crease the fold.
   - Carefully press the sticky sides of the label together, leaving a gap near the folded end.
   - Insert one end of a toothpick through the gap and stick the toothpick to the folded end of the label.

4. Repeat step 3 with the other three toothpicks and labels.

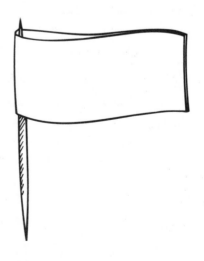

5. Use the pen to write the names of the Sun's layers on the flags: Photosphere, Convection Zone, Radiation Zone, Core.

6. Stick the flags in the model Sun as shown.

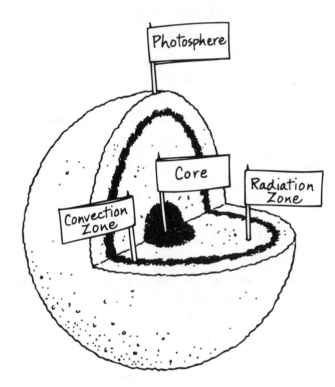

Results

A model of the layers of the Sun is made.

Why?

The center of the model Sun is the **core,** which is the hottest part of the Sun. It is in the core that **nuclear fusion** (joining of the nuclei of atoms) occurs, producing the energy of the Sun. Energy from the superhot core is slowly transmitted through the area above the core, called the **radiation zone.** From there, gas expands and rises, then cools, becomes denser, and sinks back. This circulating gas forms the **convection zone.** The next layer, called the **photosphere,** is actually the first layer of the Sun's atmosphere. But from Earth, it appears as the Sun's surface.

Janice VanCleave's Teaching the Fun of Science

Star Time

Benchmarks

By the end of grade 5, students should know that
- Earth rotates on its axis every 24 hours.

By the end of grade 8, students should know that
- Earth rotates daily on its axis and revolves yearly about the Sun.

In this investigation, students are expected to
- Demonstrate a sidereal day.
- Model the rotation and revolution of Earth.

Preparing for the Investigation

Any large ball, even a wad of yellow paper, can be used for the Sun.

Presenting the Investigation

1. Introduce the new science terms:

   axis An imaginary line through the center of an object and around which the object turns

   orbit The curved path of one body around another.

   revolve To move in a curved path around another object.

   rotate To turn on an axis.

   sidereal day The time it takes Earth to make one complete rotation of 360°.

2. Explore the new science terms:
   - A sidereal day is the time it takes Earth to rotate

once relative to the stars. This 360° rotation takes about 23 hours, 56 minutes.
- Earth rotates on its axis at a rate of about 1,000 miles (1,600 km) per hour.
- Earth revolves around the Sun at a speed of about 68,000 miles (109,000 km) per hour.
- Earth's equator is tilted at an angle of 23.50° from the perpendicular to its orbit around the Sun. More about the effects of this tilt can be found in investigation 71.

Did You Know?
- A solar day is about 4 minutes longer than a sidereal day, or 24 hours.
- An Earth year is about 365¼ solar days. Since calendars for 1 year are only 365 days, an extra day must be added every 4 years. The day is added in February and the year is called a *leap year*.

EXTENSION

The sphere, like Earth, must make slightly more than one whole turn before the paper clip points toward the Styrofoam ball again. A *solar day* is the time it takes a location on Earth to make one rotation in relation to the Sun. A solar day is the time period used for everyday affairs and is 24 hours in length. Students can demonstrate that it takes more than a 360° rotation to complete one solar day. Repeat steps 5 to 7 of the procedure, but continue rotating the clay sphere until the paper clip faces the center of the Styrofoam ball.

Star Time

PURPOSE
To demonstrate a sidereal day.

Materials

marker
sheet of typing paper
ruler
lemon-size piece of modeling clay
paper clip
pencil
4-inch (10-cm) Styrofoam ball

Procedure

1. Use the marker to draw a 6-inch (15-cm) line lengthwise across the center of the paper. Label the right end of the line "A." Four inches (10 cm) above this line and centered over it, make a second line that is 5 inches (7.5 cm) long and label the right end of the line "B."

2. Shape the clay into a sphere.

3. Insert the pencil through the center of the clay sphere from top to bottom until just the tip of the pencil sticks out the bottom.

4. Insert the paper clip in the center of one side of the sphere, so that it sticks out perpendicular to the pencil.

5. Lay the paper on the table so that the labeled ends of the line are to your right. Place the Styrofoam ball at the left end of line A.

6. Holding the pencil, position the clay sphere at the right end of line A so that the paper clip is parallel with the line and faces the center of the ball.

7. Slightly tilt the eraser end of the pencil toward the ball. Then holding the pencil, turn the sphere counterclockwise one whole turn, 360°, as you move the sphere to the right end of line B. Position the sphere so that the paper clip is parallel with line B. Observe the direction the paper clip faces in relation to the center of the ball.

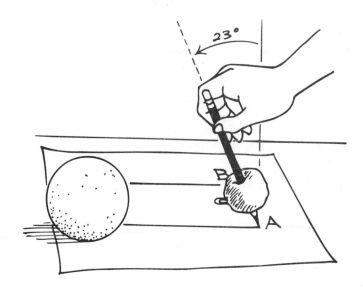

Results

At first, the paper clip faces the center of the Styrofoam ball. After the clay sphere has made one turn and has moved a portion of the way around the ball, the paper clip no longer faces center of the ball.

Why?

The pencil represents Earth's **axis** (an imaginary line through the center of an object and around which the object turns), which is tilted from the perpendicular to its orbit around the Sun. Like Earth, the clay ball **rotated** (turned on its axis) and at the same time **revolved** (moved in a curved path around another object) the way Earth moves around the Sun. Earth's curved path around the Sun is called an **orbit.** The time it takes Earth to make one complete rotation of 360° is called a **sidereal day.** After one sidereal day, a position on Earth that started out facing the center of the Sun does not end up that way, because while Earth was rotating, it was also revolving to another point in its orbit.

Rotate

Benchmarks

By the end of grade 5, students should know that

- To people on Earth, the rotation of Earth makes it seem as though the Sun is orbiting Earth.

By the end of grade 8, students should know that

- Earth's axis is tilted relative to its orbit about the Sun. This tilt makes the Sun appear at different altitudes during the year.

In this investigation, students are expected to

- Demonstrate how Earth's rotation makes the Sun appear to move across the sky.

Presenting the Investigation

1. Introduce the new science terms:

   horizon An imaginary line where the sky appears to meet Earth's surface.

   Northern Hemisphere The region above, or north of, Earth's equator.

   North Pole The north end of Earth's axis.

   Southern Hemisphere The region below, or south of, Earth's equator.

   South Pole The south end of Earth's axis.

2. Explore the new science terms:
   - The Sun appears to move across the sky because Earth rotates.
   - Earth rotates from west to east.
   - The Sun appears to move from the eastern horizon to the western horizon each day.
   - The Sun rises above the eastern horizon and sets below the western horizon.
   - As viewed from above the North Pole, Earth appears to rotate in a counterclockwise direction.

- From above the South Pole, Earth appears to rotate clockwise.

Did You Know?

The Sun gives off *white light*, which is made up of all the rainbow colors. The Sun looks yellow to an observer on Earth because as the Sun's white light passes through Earth's atmosphere, particles of matter and air molecules scatter the blue light, which makes the sky appear blue. White light minus blue light produces yellow light.

- The equator is an imaginary line that divides Earth into northern and southern halves. It is perpendicular to Earth's axis.
- In the Northern Hemisphere, the Sun appears to move across the southern sky.
- In the Southern Hemisphere, the Sun appears to move across the northern sky.
- Earth's equator is tilted at an angle of 23.50° from the perpendicular to its orbit around the Sun. More about the effects of this tilt can be found in investigation 71.

EXTENSION

Because of Earth's revolution around the Sun, the Sun has an apparently eastward yearly journey across the sky. Because of the tilt of Earth's axis relative to its orbit around the Sun, the Sun's altitude at midday varies during the year, and the Sun does not rise and set day after day in the same place. Students can discover these changes using a shadow in sunlight. *Caution students not to look directly at the Sun because it can permanently damage their eyes.* A safe way to discover the apparent changes in the Sun's position is by observing the changing position of the shadow of a stationary object, such as a flagpole, at the same time each day for an extended period.

70 Rotate

PURPOSE

To determine why the Sun appears to move across the sky.

Materials

pencil
clay sphere from investigation 69, "Star Time"
crayons
2 index cards
ruler
2 grape-size balls of modeling clay

Procedure

1. Use the pencil to draw a line perpendicular to the pencil around the middle of the clay sphere (the model Earth).
2. Move the paper clip to a spot in the clay above the line to represent an observer in the Northern Hemisphere.
3. Use the crayons to draw a symbol of the Sun on one of the index cards. Label the right side of the Sun "West" and the left side "East."
4. Draw six or more stars on the other index card.
5. Stand the index cards 12 inches (30 cm) apart in the grape-size balls of clay so that the Sun and the stars face each other.
6. Stand the model Earth between the index cards so that the paper clip observer faces the star card. Slightly tilt the model toward the Sun.
7. Slowly rotate the model Earth counterclockwise until the paper clip observer faces the right edge (west side) of the Sun card.
8. Continue rotating the model Earth counterclockwise until the paper clip observer faces the left edge (east side) of the Sun card.

Results

As the Earth model rotates away from the stars, the paper clip observer on Earth faces the west side of the Sun first and the east side last.

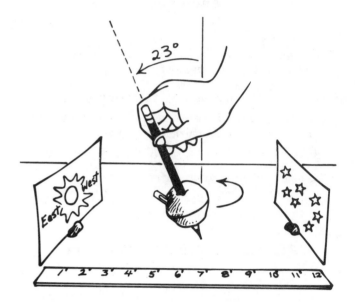

Why?

The line around the model Earth represents the **equator,** an imaginary line that divides Earth into northern and southern halves. The equator is midway between Earth's poles. The ends of Earth's axis, represented by the pencil, are the **North Pole** (the north end of Earth's axis) and the **South Pole** (the south end of Earth's axis). The region above, or north of, the line is called the **Northern Hemisphere,** and the region below, or south of, the line is called the **Southern Hemisphere.**

In the Northern Hemisphere, the Sun appears to rise above the eastern **horizon** (an imaginary line where the sky appears to meet Earth), move across the southern sky, and set below the western horizon. If you could view Earth from above the North Pole, you would see Earth, like the model, rotating counterclockwise. The paper clip observer on the clay sphere first sees the western side of the Sun diagram, then the eastern side comes into view as the sphere rotates. Since we are moving with Earth as it rotates, it appears that the Sun is moving across the sky from east to west, but it is actually Earth that is moving from west to east.

Tilted

Benchmarks

By the end of grade 5, students should know that

- Earth revolves about the sun every year.

By the end of grade 8, students should know that

- Earth's axis is tilted relative to the plane of Earth's yearly orbit around the Sun. As Earth revolves around the Sun, sunlight falls more intensely on different parts of Earth during the year, causing the seasons.

In this investigation, students are expected to

- Identify how the position of Earth in relation to the Sun results in reversed seasons in the Northern and Southern Hemispheres.
- Model the way in which Earth's motion produces seasons.
- Identify the position of Earth in relation to the Sun during different seasons.

Presenting the Investigation

1. Introduce the new science terms:

   summer solstice The day when Earth's North Pole is tilted closest to the Sun, on or about June 21 in the Northern Hemisphere.

   winter solstice The day when Earth's South Pole is tilted farthest away from the Sun, on or about December 22 in the Northern Hemisphere.

2. Explore the new science terms:
   - Earth's orbit around the Sun is about 599 million miles (958 million km).
   - During Earth's revolution around the Sun, the ends of Earth's axis (North and South Poles) are tilted toward the Sun for part of the year and away from the Sun for the other part of the year.

- Earth's tilt changes the concentration of the Sun's rays that reach certain regions of Earth and the number of hours of sunlight each day.
- Increased sunlight and direct rays cause an increase in temperature.
- In many regions of Earth, there are four seasons: winter, spring, summer, and autumn. Other regions, such as at the equator, have very little change over the course of the year.
- Many seasons are identified by temperature, but in some regions of Earth, the seasons differ dramatically in the amount of rain received, so there is a dry and a rainy season.
- The axis of a celestial body above its orbital plane is called the *north pole*.

Did You Know?

The distance of Earth from the Sun doesn't cause seasons. When Earth is farthest from the Sun, it is summer in the Northern Hemisphere but winter in the Southern Hemisphere.

EXTENSION

Ask students to research Earth's orbit around the Sun. Is every point on Earth's orbit an equal distance from the Sun? What is perihelion? What is aphelion? (The distance between Earth and the Sun changes during Earth's orbit around the Sun. *Perihelion* is the point of Earth's orbit closest to the Sun. At perihelion, Earth is about 95 million miles [152 million km] from the Sun. *Aphelion* is the point of Earth's orbit farthest from the Sun. At aphelion, Earth is about 92 million miles [147 million km] from the Sun.)

Tilted

PURPOSE

To determine why seasons are reversed in the Northern and Southern Hemispheres.

Materials

flashlight
model Earth from investigation 70, "Rotate"
protractor

Procedure

1. Hold the flashlight about 6 inches (15 cm) from the model Earth. The flashlight represents the Sun.

2. Use the protractor to measure as you tilt the pointed end of the pencil (the North Pole) about 23° toward the flashlight. Observe the area of the sphere that is lit up by the flashlight and make a diagram in row 1 of the Seasons Data table. In your diagram, shade in the area of the sphere that is not lit up by the flashlight and leave the lit area unshaded.

3. Tilt the eraser end of the pencil (the South Pole) about 23° toward the flashlight. Observe the area of the sphere that is lit up and make a diagram in row 2.

Results

When the top of the sphere is tilted toward the flashlight, more light hits the top half of the sphere than the bottom. The reverse is true when

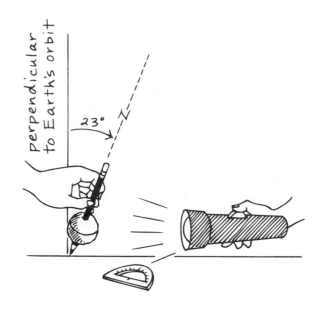

the top of the sphere is tilted away from the flashlight.

Why?

Earth's equator is tilted at an angle of 23.5° from the perpendicular to its orbit around the Sun. This means that during part of its orbit, the North Pole is tilted toward the Sun, and during the other part the North Pole is tilted away from the Sun. The Northern Hemisphere receives the most solar energy when the North Pole is tilted toward the Sun. The day of the year when Earth's North Pole is tilted closest to the Sun is called the **summer solstice** and occurs on or about June 21 in the Northern Hemisphere. The summer season begins on this day. The day when Earth's North Pole is tilted farthest away from the Sun is called the **winter solstice** and occurs on or about December 22. The winter season begins on this day. The dates of the summer and winter solstices are reversed in the Southern Hemisphere.

SEASONS DATA	
Earth Position	**Diagram**
North Pole tilted toward the Sun	
South Pole tilted toward the sun	

Janice VanCleave's Teaching the Fun of Science

Same Face

Benchmarks

By the end of grade 5, students should know that

- The features of the Moon facing Earth always look the same.

By the end of grade 8, students should know that

- The Moon orbits Earth once in about every 27 days.

In this investigation, students are expected to

- Describe the motion of the Moon in relation to Earth.
- Identify gravity as the force that keeps the Moon in its orbit around Earth.
- Model the period of rotation and the period of revolution of the Moon.

Preparing for the Investigation

You may wish to supply large and small lids instead of a drawing compass to assist students in drawing the large circle representing the Moon's orbit and the small circle representing Earth.

Presenting the Investigation

1. Introduce the new science terms:

   period of revolution The time it takes one body to revolve around another body.

   period of rotation The time it takes a body to rotate once on its axis.

2. Explore the new science terms:

Did You Know?

On October 4, 1959, the Soviet Union's space probe *Lunik 3* sent the first pictures of the far side of the Moon back to Earth. About 70 percent of the lunar far side was photographed. No one had ever seen the far side before.

- The Moon's period of revolution around Earth is 27.3 days.
- The Moon's period of rotation is 27.3 days.
- The force of gravity not only keeps the Moon in its orbit around Earth, but also affects the rate of the Moon's rotation so that the Moon always keeps the same face to Earth as it revolves around Earth.

EXTENSION

Ask students to research the Moon's orbital motion. What is the Moon's average speed? What is its sidereal period? What is its perigee? What is its apogee? (The Moon's average speed is 2,295 miles [3,672 km] per hour. Its sidereal period is a sidereal month, which is the time it takes the Moon to orbit Earth once—27.3 days. *Perigee* is the point in its orbit when the Moon is closest to Earth—222,750 miles [356,400 km]. *Apogee* is the point in its orbit when the Moon is farthest from Earth—254,188 miles [406,700 km].)

Same Face

PURPOSE

To model a relation between the period of revolution and the period of rotation of the Moon.

Materials

drawing compass
pencil
sheet of white copy paper
¾-inch (1.9-cm) round color-coding label, any color

Procedure

1. Use the compass to draw as large a circle as possible on the paper.
2. In the center of the large circle, draw a smaller circle and label it "Earth."
3. Stick the round color-coding label just above the pointed end of the pencil.
4. Stand the pencil upright with its point on the outer circle and the label facing Earth as shown.
5. Move the pencil around the circle once. During the revolution, rotate the pencil so that the label faces Earth at all times. Count the number of times the pencil rotates.

Results

The pencil rotates one time.

Why?

The pencil represents the Moon, and the label the side of the Moon that faces Earth. Earth's gravitational pull on the Moon affects the rate of the Moon's rotation. The Moon rotates once during each revolution around Earth. Therefore, like the label in this experiment, the same side of the Moon faces Earth at all times. The Moon's **period of rotation** (the time it takes to rotate once on its axis) and **period of revolution** (the time it takes to orbit Earth once) are both about 27.3 days.

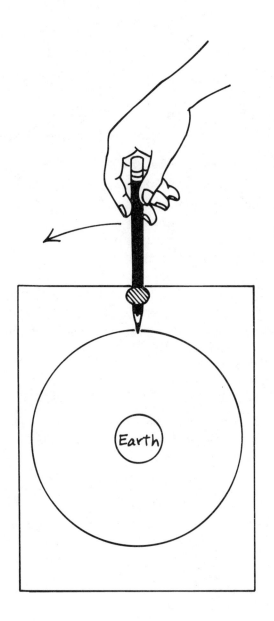

Moon Watch

Benchmarks

By the end of grade 5, students should know that

- The Moon looks a little different every day, but looks the same again about every 4 weeks.

By the end of grade 8, students should know that

- The Moon's orbit around Earth determines which part of the Moon is lighted by the Sun and how much of that part can be seen from Earth, causing what is called phases of the Moon.

In this investigation, students are expected to

- Identify and observe the phases of the Moon.

Preparing for the Investigation

Students will make individual observations, but can work in groups to prepare their observation data table—a calendar. For each group, provide one copy of the calendar month(s) needed for the observation period. Caution students not to observe the Moon about 2 days before and 2 days after the new moon. You can mark these days on their prepared observation calendar.

Presenting the Investigation

1. Introduce the new science terms:

   crescent moon The phase of the Moon in which the lighted area of the side of the Moon facing Earth resembles a ring segment with pointed ends.

   first quarter moon The phase of the Moon that follows the new moon in which half of the side of the Moon facing Earth is lighted.

   full moon The phase of the Moon in which the side of the Moon facing Earth is fully lighted.

   gibbous moon The phase of the Moon, occurring before and after full moon, in which more than half of the side of the Moon facing Earth is lighted.

   new moon The phase of the Moon in which the side of the Moon facing Earth is not lighted.

Did You Know?

A *blue moon* has nothing to do with color. It is the second full moon in a month that occurs every three years or so.

   phases of the moon The apparent changes in size and shape of the side of the Moon facing Earth that is lighted by the Sun.

   third quarter moon The phase of the Moon in which half of the side of the Moon facing Earth is lighted.

   wane To get smaller.

   wax To get larger.

2. Explore the new science terms:

   - About half of the Moon's surface is lighted by the Sun. As the Moon orbits Earth, the amount and orientation of the lighted surface that is visible from Earth changes. This causes the different phases of the moon.

   - It takes the Moon about 27.3 days to revolve around Earth, but because Earth is revolving around the Sun, it takes the Moon about 29.5 days to return to the same shape observed on the first night of observation.

   - The time it takes the Moon to complete its phases from new moon to new moon is called a *lunar synodic month*.

   - Waxing phases are from new moon to full moon. The Moon appears to get larger.

   - Waning phases are from full moon to new moon. (In the Northern Hemisphere, the Moon is lighted on the left when it is waning. You can look at the Moon and determine whether it is waxing or waning by whether the right or the left is lighted.)

EXTENSION

Students can use the approximate times in the Moonrise and Moonset table shown here as a guide to determine when to look for the Moon. The class can provide accurate times of moonrise and moonset during the time the phases are being observed.

Approximate Times of Moonrise and Moonset			
Shape	**Phase of the Moon**	**Moonrise**	**Moonset**
●	new	dawn	sunset
◑	first quarter	noon	midnight
○	full	sunset	dawn
◐	third quarter	midnight	noon

Moon Watch

PURPOSE

To observe the phases of the Moon.

Materials

sheet of typing paper
pen
ruler
calendar that gives phases of the Moon

Procedure

1. Use the paper, the pen, and the ruler to draw a 5-week calendar.

2. Fill in the dates on the calendar, starting with the day you prepare the calendar. Note that the calendar may include parts of two months.

3. Observe the shape of the Moon for 29 days. Look for the Moon during the day as well as at night. Draw the shape of the Moon for each day on the calendar. Label daytime drawings "D" and nighttime drawings "N." *NOTE: Make no observation for at least 2 days before and 2 days after new moon—when the side of the Moon facing Earth is dark. The new moon is close to the Sun and you could damage your eyes if you look at the Sun.*

4. Study the different phases shown here and use them to label the drawings on your calendar.

Results

It takes about 29 days for the Moon to return to the same shape observed on the first night of observation.

Why?

Half of the Moon is always lighted by the Sun, but only one side of the Moon faces Earth. The different shapes of the lighted areas are called **phases of the Moon.** The **new moon** has no lighted areas in the side facing Earth. After new moon, the lighted area of the Moon **waxes** (gets larger) until **full moon,** when the side of the Moon facing Earth is fully lighted. Between new moon and full moon are other phases, including **crescent moon** (when the lighted area of the side facing Earth resembles a ring segment with pointed ends), **first quarter moon** (when half of the side facing Earth is lighted), and **gibbous moon** (when more than half of the side facing Earth is lighted). In the Northern Hemisphere, these phases appear on the right side of the Moon. After full moon, the lighted area of the Moon **wanes** (gets smaller) until new moon. Between full moon and new moon, the phases reverse. The opposite side of the Moon—left side—is lighted. Instead of first quarter, the phase when half of the side facing Earth is lighted is called **third quarter moon.** This cycle from one new moon to the next takes about 29 days. The lighted areas of the Moon during the different phases are reversed in the Southern Hemisphere.

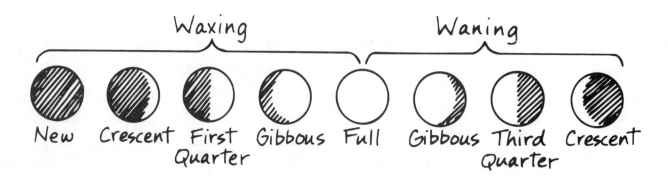

Shadows

Benchmarks

By the end of grade 5, students should know that

- To an observer on Earth, Earth's rotation on its axis makes it appear that the Sun and the Moon move across the sky each day.

By the end of grade 8, students should know that

- The Moon orbits Earth once in about every 27 days. At times, but not each month, the Moon moves between Earth and the Sun and blocks the Sun's light.

In this investigation, students are expected to

- Model a solar eclipse.
- Describe the positions of the Sun, Earth, and the Moon during a solar eclipse.

Presenting the Investigation

1. Introduce the new science terms:

   apparent size The size a distant object appears to be.

   eclipse To cause an eclipse, an event in which one celestial body passes in front of and blocks the light of another.

   elliptical Oval-shaped.

   solar eclipse An eclipse in which the Moon passes in front of and blocks the light of the Sun.

   total solar eclipse A solar eclipse in which the Moon blocks all the Sun's light.

2. Explore the new science terms:

   - A *shadow* is a dark shape cast upon a surface when something blocks light. The dark inner part of a

   shadow is called the *umbra,* and the lighter outer region is called the *penumbra.*

   - Because Earth rotates, the shadow of the Moon sweeps across Earth's surface during a total solar eclipse. The Moon's shadow is about 188 miles (300 km) wide.

   - Since the size of the Moon's shadow during a total solar eclipse is only about 188 miles (300 km) wide, a total solar eclipse occurs rarely for any one spot on Earth.

Did You Know?

During a total solar eclipse, only part of the side of Earth that faces the Sun and the Moon is in the shadow of the Moon. Observers in the umbra see a total eclipse, while those in the penumbra see a *partial solar eclipse* (when the Moon is not quite in a direct enough line with the Sun and Earth to block all the light of the Sun from an observer on Earth). Those outside the Moon's shadow see no eclipse.

EXTENSION

When the Moon is far enough from Earth to appear slightly smaller than the Sun, the Moon does not completely eclipse the Sun. An outer ring of the Sun's *photosphere* (the bright outer layer of the Sun) is visible. This event is called an *annular eclipse.* Have students demonstrate an annular eclipse by repeating the experiment, slowly moving the clay ball away from the face until only a small outer ring of the Styrofoam ball is visible around the clay ball.

Shadows

PURPOSE

To model how the Moon can hide the Sun.

Materials

grape-size ball of clay
2 sharpened pencils
3-inch (7.5-cm) Styrofoam ball

Procedure

1. Stick the ball of clay on the point of one of the pencils and the Styrofoam ball on the other pencil's point.

2. Hold the pencil with the Styrofoam ball in front of your face at arm's length.

3. Close one eye and hold the pencil with the clay ball so that the ball is in front of but not touching your open eye. Slowly move the clay ball away from your face toward the Styrofoam ball. As you move the clay ball, observe how much of the Styrofoam ball is hidden by the clay ball at different distances.

Results

The closer the clay ball is to your face, the more it hides the Styrofoam ball.

Why?

The closer an object is to your eye, the bigger its **apparent size** (the size an object at a distance appears to be). The small ball of clay can totally hide the larger Styrofoam ball, blocking it from view. In the same way, the Moon, with a diameter of 2,173 miles (3,476 km), can sometimes hide and block the light of the much larger Sun, which has a diameter of 870,000 miles (1,392,000 km).

When the Moon passes directly between the Sun and Earth, and all three are in a straight line, the Moon **eclipses** (passes in front of and blocks the light of) the Sun. In this position, observers on Earth see a **solar eclipse.** The Sun is about 400 times larger than the Moon, but at times during the Moon's **elliptical** (oval-shaped) orbit, the Moon is about 400 times nearer Earth. It is in this position that the Moon and the Sun appear to be the same size. In a solar eclipse in which the Moon appears as large as the Sun, the Moon completely blocks the light of the Sun. This event is called a **total solar eclipse.**

Appendix 1:
Graduated Cylinder

PURPOSE

To make a model of a graduated cylinder.

Materials

crayon—red or any dark color
graduated cylinder pattern
laminating material (optional)
scissors
ruler
pen
transparent tape

Procedure

1. Color the liquid strip portion of the graduated cylinder pattern.

2. Laminate the colored patterns. This is an optional step.

3. Cut out the four areas indicated.

4. Cut along the dotted line to separate the liquid strip from the other sections.

5. Score each fold line by laying the ruler along each of the fold lines, then trace the lines with the pen.

6. Fold the paper along fold line 1, then along fold line 2, and secure the folded sections together with tape.

7. Insert the curved end of the liquid strip in the liquid strip slot so that the colored side of the strip is visible through the openings in the front of the model.

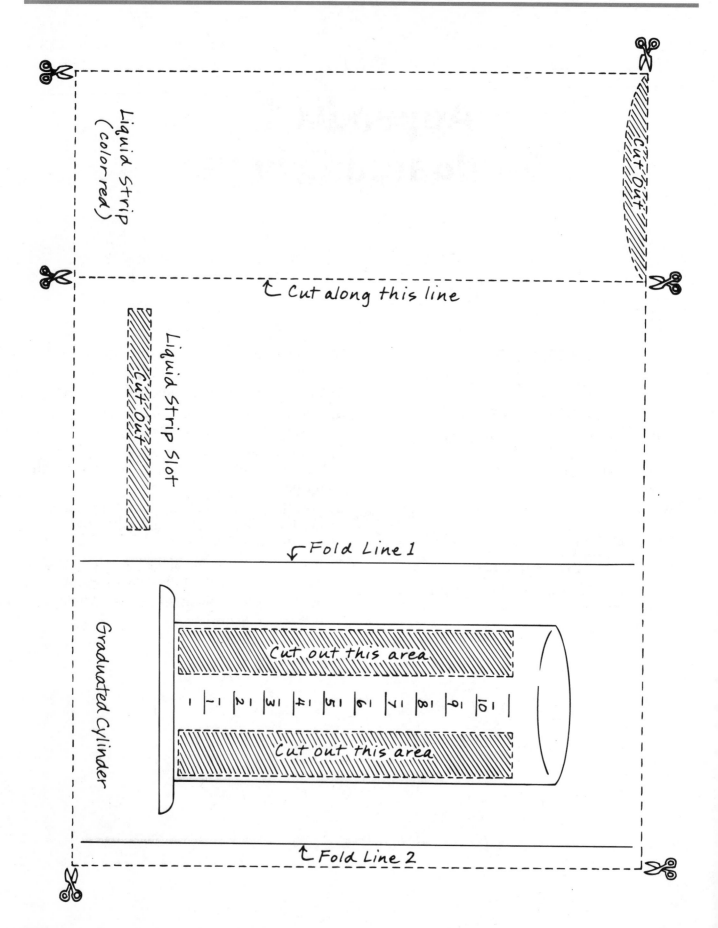

Liquid Strip
(color red)

Cut Out

↳ Cut along this line

Liquid Strip Slot

Cut Out

↳ Fold Line 1

Graduated Cylinder

Cut out this area

| 1 | 2 | 3 | 4 | 5 | 6 | 7 | 8 | 9 | 10 | |

Cut out this area

↳ Fold Line 2

Appendix 2: Thermometer

PURPOSE

To make a model of a thermometer.

Materials

red crayon
thermometer pattern
laminating material (optional)
scissors
ruler
pen
transparent tape

Procedure

1. Color the liquid strip and bulb portion of the thermometer pattern red.

2. Laminate the colored patterns. This is an optional step.

3. Cut out the two indicated areas.

4. Cut along the dotted line to separate the liquid strip from the other sections.

5. Score each fold line by laying the ruler along each of the fold lines, then trace the lines with the pen.

6. Fold the paper along fold line 1, then along fold line 2, and secure the folded sections together with tape.

7. Insert the liquid strip in the liquid strip slot so that the colored side of the strip is visible through the openings in the front of the model.

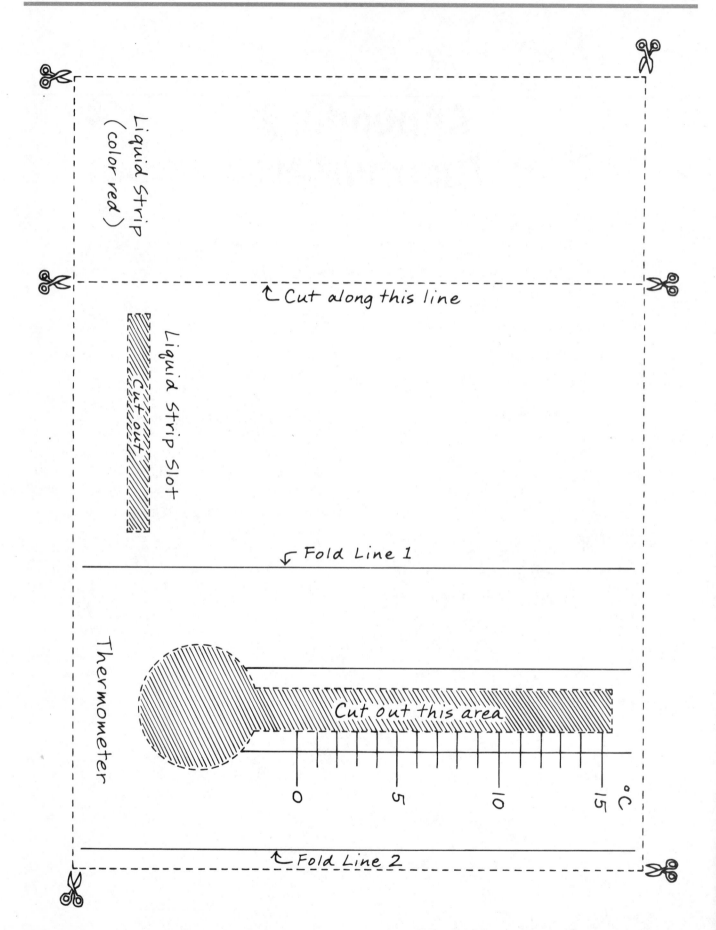

Liquid Strip
(color red)

↰ Cut along this line

Liquid Strip Slot

Cut out

Fold Line 1 ↰

Thermometer

Cut out this area

0

5

10

15

°C

↑ Fold Line 2

192

Appendix 3:
Sources of Scientific Supplies

CATALOG SUPPLIERS

The following science supply companies carry materials such as paramecia and other live specimens, slides, and student thermometers that can be purchased by mail order.

Carolina Biological Supply Company
2700 York Road
Burlington, NC 27215
(800) 334-5551

Cuisenaire
10 Bank Street
P.O. Box 5026
White Plains, NY 10606
(800) 237-3142

Delta Education, Inc.
P.O. Box 915
Hudson, NH 03051-0915
(800) 258-1302

Fisher Scientific
Educational Materials Division
485 South Frontage Road
Burr Ridge, IL 60521
(708) 655-4410
(800) 766-7000

Frey Scientific Division of Beckley Cardy
100 Paragon Parkway
Mansfield, OH 44903
(800) 225-3739

NASCO
901 Janesville Avenue
P.O. Box 901
Fort Atkinson, WI 53538
(800) 677-2960

Sargent-Welch
911 Commerce Court
Buffalo Grove, IL 60089
(800) 727-4368

Showboard
P.O. Box 10656
Tampa, FL 33679-0656
(800) 323-9189

Ward's Natural Science
5100 West Henrietta Road
Rochester, NY 14586
(800) 962-2660

SOURCES OF ROCKS AND MINERALS

The following stores carry rocks and minerals and are located in many areas. To find the stores near you, call the home offices listed below.

Mineral of the Month Club
1290 Ellis Avenue
Cambria, CA 93428
(800) 941-5594
cambriaman@thegrid.net
www.mineralofthemonthclub.com

Nature Company
750 Hearst Avenue
Berkeley, CA 94701
(800) 227-1114

Nature of Things
10700 West Venture Drive
Franklin, WI 53132-2804
(800) 283-2921

The Discovery Store
15046 Beltway Drive
Dallas, TX 75244
(214) 490-8299

World of Science
900 Jefferson Road
Building 4
Rochester, NY 14623
(716) 475-0100

Glossary

absorb   (1) To soak up. (2) To take in. (3) To receive sound without echo.

active layer   The thin layer of ground above permafrost that freezes in the winter and thaws in the summer.

adaptation   An adjustment in physical characteristics or behavior that allows a species to survive in the conditions of its environment.

agent of erosion   A natural force, such as water, wind, ice, or gravity, that transports eroded materials.

air mass   A large body of air that is about the same temperature throughout.

allele   One of several different forms of a specific gene.

alpine tundra   The tundra of a region above the tree line at high altitudes.

angular distance   The apparent distance between two celestial bodies, measured in degrees.

apparent size   The size a distant object appears to be.

Arctic region   The region of Earth near the North Pole; the area above latitude 66½° N.

Arctic tundra   The tundra of the Arctic region.

arteries   Large blood vessels that carry red oxygen-rich blood from the heart.

arterioles   Ends of arteries that connect to capillaries.

asexual reproduction   Reproduction in which there is only one parent and the offspring are identical to the parent.

atmosphere   The blanket of gases surrounding a celestial body.

atom   The smallest unit of an element; a building block of matter.

attraction   The ability to be drawn toward.

auxin   A plant hormone that causes changes in the growth of a cell.

average speed   The total distance traveled divided by the total time.

axis   An imaginary line through the center of an object and around which the object turns.

balanced force   Force applied equally to an object from opposite directions.

ballooning   A technique that spiderlings use to move to different places.

beach   A shore with a smooth stretch of sand or pebbles.

behavior   An observable response in an organism.

biome   A large ecosystem characterized by the plants that occur there due to the climate of the region.

blood   A fluid that carries materials throughout the body.

bond   A force that links atoms together.

camouflage   Concealment by protective coloration that helps an animal blend in with its surroundings.

canopy   The umbrellalike top layer of a tropical forest that is formed by the tops of tall broad-leaved evergreen trees.

capillaries   Microscopic blood vessels that link arteries and veins.

capillary links   A name used for arterioles and venules.

celestial bodies   Natural objects in the sky, such as planets, moons, stars, and suns.

cell   A building block of living things.

cell membrane   The thin outer skin that holds a cell together and allows materials to move into and out of the cell.

Celsius scale   A temperature scale in which the freezing point of water is 0° and the boiling point is 100°.

cementation   The binding together of materials.

cerebellum   The part of the brain that controls muscle action.

cerebrum   The largest part of the brain, which controls thoughts.

characteristics   Natural features.

chemical properties   Characteristics that describe the behavior of a substance when its identity is changed.

chemical reaction   A process by which atoms interact to form one or more new substances.

chemical weathering   A type of weathering that changes the chemical properties of crustal materials.

chlorophyll   A green pigment in plant cells that absorbs the light needed for photosynthesis.

chromatography   A method of separating a mixture into its different substances.

chromosome   A rod-shaped structure in the nucleus of a cell that contains DNA.

cilia   Tiny hairlike parts used by some unicellular organisms for locomotion.

circulatory system   A closed network of blood vessels through which blood flows in the body.

classification   The arranging of organisms into groups based on the similarities of their characteristics.

clastic rock   A type of sedimentary rock formed when sediments from preexisting rock are compacted and cemented together.

cold-blooded animal   An animal whose internal body temperature changes with the temperature outside its body.

cold desert   A desert having temperatures that are below freezing for part of the year.

compaction   The squeezing together of materials.

compound   A substance made of molecules that are alike.

condensation   The phase change from a gas to a liquid.

conduction   The transfer of heat from one particle to another by collision.

cone   The reproductive structure of a conifer.

conifer   A tree or shrub whose seeds are stored in cones and that usually has needle-shaped leaves.

coniferous forest   A forest made up mainly of conifers and lying below the tree line.

conifers   Cone-bearing trees.

contract   To draw together.

convection   The transfer of heat from one region to another by the circulation of currents in a fluid.

convection current   The circular movement of fluids of unequal temperature.

convection zone   The layer of the Sun between the radiation zone and the photosphere.

core   The center of a celestial body. Earth's core is the part of the geosphere below the mantle and made up mostly of two metallic elements, iron and nickel. The Sun's core is the hottest part of the Sun.

cotyledon   A simple leaf beneath a seed coat that stores food for a developing plant.

country rock   The common rock of a region.

crescent moon   The phase of the Moon in which the lighted area of the side of the Moon facing Earth resembles a ring segment with pointed ends.

cross-link   A chemical bridge between polymer molecules.

crust   The outer layer of Earth's geosphere, on which organisms live.

crystal   A solid with flat surfaces that has particles arranged in repeating patterns.

cutting   A piece cut from a plant to grow a new plant.

cytoplasm   A clear jellylike material occupying the region between the nucleus and the cell membrane of a cell that contains substances and particles that work together to sustain life.

deciduous   Having leaves that are lost, usually in autumn.

deciduous forest   See **temperate forest.**

degree   A unit for measuring angles.

density   The mass per unit volume of a substance.

deoxyribonucleic acid (DNA)   Chemical molecules in chromosomes that control cell activity and determine hereditary traits.

desert   A biome that receives less than 10 inches (25 cm) of rain per year.

diatomic molecules   A molecule made up of two atoms of the same kind.

diffuse   To spread freely and become evenly distributed.

dissolve   To break up and thoroughly mix with another substance, as salt in water.

divergent boundary   A border where tectonic plates separate and new crustal material is added.

dominant allele   A gene form that when present determines the trait.

eclipse   To cause an eclipse, an event in which one celestial body passes in front of and blocks the light of another.

ecosystem   A region where living and nonliving interact with each other and their environment.

egg   A female sex cell.

elasticity   The physical property of being able to return to the original length or shape after being stretched.

electromagnetic wave   A disturbance in electric and magnetic fields; disturbance that can move through space.

electron   A negatively charged particle outside the nucleus of an atom.

element   A substance made up of atoms that are alike.

elliptical   Oval-shaped.

embryo   An organism in the earliest stage of its development, such as the immature plant inside a seed.

energy (E)   The ability to do work.

environment   The conditions around organisms that affect their lives, including weather, land, and food.

epicotyl   The part of a plant embryo above the cotyledons' point of attachment that develops into a plant's stem, leaves, flowers, and fruit.

equator   An imaginary line around Earth at 0° latitude that divides Earth into northern and southern halves.

erosion   The process by which rock and other materials in Earth's crust are broken down and carried away by agents of erosion.

evaporate   To change from a liquid to a gas.

evaporation   The phase change from a liquid to a gas.

evergreen   Having leaves that are not lost and stay green year-round.

exosphere   The highest outermost layer of Earth's atmosphere, starting at the thermosphere.

expand   To move farther apart.

extinct   No longer in existence.

fertilization   The joining of two sex cells, an egg and a sperm, from two parents.

first quarter moon   The phase of the Moon that follows the new moon in which half of the side of the Moon facing Earth is lighted.

fluid   A material that flows: gas or liquid.

force   A push or a pull on matter.

forest   A biome that contains a large group of trees growing close together with various kinds of smaller plants.

formula   A symbolic representation of a molecule.

friction   A force that opposes the motion of one object whose surface is in contact with another object.

front   A boundary between cold and warm air masses.

full moon   The phase of the Moon in which the side of the Moon facing Earth is fully lighted.

gas   A substance in a phase of matter characterized by no definite shape or volume.

gender   The sex of an organism: male or female.

gene   The part of a chromosome that determines hereditary traits; consists of DNA.

gene site   The site where a gene is positioned on a chromosome.

genotype   The genetic makeup of an organism or group of organisms as determined by alleles.

geosphere   Solid part of Earth; crust, mantle, core.

geotropism   See **gravitropism**.

germination   The process by which a seed begins to grow.

gibbous moon   The phase of the Moon, occurring before and after full moon, in which more than half of the side of the Moon facing Earth is lighted.

glacier   A large body of land ice that flows slowly downhill.

glide   To move smoothly and effortlessly.

graduated cylinder   An instrument used to measure volume.

grassland   A semiarid biome whose vegetation is mostly grass with few trees or shrubs.

gravitational potential energy (GPE)   Potential energy due to an object's height above a surface.

gravitropism   Growth or movement of a plant in response to gravity; also called **geotropism**.

gravity   A force of attraction between all objects in the universe.

gravity rate (G.R.)   The surface gravity of a celestial body divided by the surface gravity of Earth, thus Earth's G.R. equals 1.

greenhouse effect   The warming of Earth by gases in the atmosphere, which trap infrared radiation from the Sun and reradiate it toward Earth, just as the glass or plastic of a greenhouse traps and reradiates infrared radiation within it.

grip   A tight hold or firm grasp on an object.

halite   The mineral form of table salt, made of sodium chloride crystals.

hand-eye coordination   The ability to move your hand in response to what your eyes see.

heat   A form of energy that is related to the total kinetic energy of all the particles of a material, and that is transferred from a warm material to a cool material because of their temperature differences.

heredity   The passing of traits from one generation to the next.

heterogeneous   Not visibly the same throughout.

homogeneous   Visibly the same throughout.

horizon   An imaginary line where the sky appears to meet Earth's.

humidity   Dampness of the air.

hybrid   An offspring whose alleles are different for a trait.

hypocotyl   The part of a plant embryo beneath the cotyledons' point of attachment, the upper part of which develops into the plant's shoot system.

iceberg   A piece of a glacier that has broken away and floats in the ocean.

igneous rock   Rock formed when magma cools and solidifies.

inertia   The tendency of an object to remain at rest or to resist any change in its state of motion unless acted on by an outside force.

infrared radiation   Radiation that all objects give off and that produces heat when absorbed.

inherit   To receive traits from parents.

innate behavior   An inherited response; not learned.

insulator   A material that is a poor conductor of heat.

International System of Units (SI)   The measurement system used primarily in science and technology around the world. Commonly called the **metric system**.

joule (J)   An SI unit by which work is measured.

kinetic energy (KE)   Energy that a moving object possesses because of its motion.

latitude   Distance in degrees north or south of the equator.

lava   Molten rock that has reached Earth's surface.

law of conservation of energy   A law of physics that states that energy can be changed from one kind to another, but cannot be created or destroyed under normal conditions.

law of conservation of mechanical energy   A law of physics that states that the sum of the potential and kinetic energy of an object remains the same as long as no outside force acts on it.

learned behavior   A response that is acquired by an organism's experience.

lift   An upward force on a flying object.

liquid   A substance in a phase of matter characterized by a definite volume but no definite shape.

liter (L)   The SI or metric unit of volume.

lithification   The hardening of sediments into rock.

lithosphere   The part of Earth consisting of the crust and the upper part of the mantle.

locomotion   The act of moving from one place to another.

magma   Molten rock beneath Earth's crust.

magnet   An object that is surrounded by a magnetic field and attracts magnetic materials.

magnetic field   The space around a magnet where a magnetic force can be detected.

magnetic force   The attraction between magnets or between a magnet and a magnetic material.

magnetic lines of force   A pattern of lines representing the magnetic field around a magnet.

magnetic materials   Materials that can be attracted to or magnetized by a magnet, such as iron and steel.

magnetic poles   One of two ends of a magnet where the magnetic field is strongest.

magnetism   Magnetic force.

mammal   An animal that has hair and feeds its young on milk.

mantle   The layer of Earth's geosphere between the core and the crust, which is made up mostly of silicates.

mass   An amount of material.

matter   Anything that occupies space and has mass; the stuff the universe is made of.

mechanical energy   Energy of motion; sum of the potential and kinetic energy of an object.

mechanical weathering   A type of weathering that breaks down crustal materials by physical means.

meiosis   The process of cell division by which sex cells are produced.

melt To change from a solid to a liquid.

melting point The temperature at which a solid changes to a liquid.

meniscus The curved upper surface of a column of liquid.

mesosphere The layer of Earth's atmosphere between the stratosphere and the thermosphere.

metamorphic rock Rock that forms from other types of rock by pressure and heat within Earth's crust.

metamorphism The process by which heat and pressure change the makeup, texture, or structure of rocks.

metric system See **International System of Units (SI).**

midocean ridge One of a number of ridges forming a continuous chain of underwater mountains around Earth.

milliliter (ml) One-thousandth of a liter.

mineral A solid found in Earth's crust that makes up rocks.

mixture A combination of two or more substances. Mixtures may be heterogeneous or homogeneous.

molecule A group of two or more atoms held together by bonds.

molten Melted.

motion The act or process of changing position.

muscular force A force caused by changes in the length of muscles.

negative gravitropism Upward growth or movement of a plant, in the direction opposite the force of gravity.

new moon The phase of the Moon in which the side of the Moon facing Earth is not lighted.

newton (N) The SI unit of weight.

Northern Hemisphere The region above, or north of, Earth's equator.

North Pole The north end of Earth's axis.

nuclear fusion Joining of the nuclei of atoms.

nucleus (plural **nuclei**) (1) The center of an atom. (2) A spherical or oval-shaped body in a cell that controls cell activity.

orbit The curved path of one body around another.

organism A living thing.

outer ear The visible outer part of the ear that collects sound waves and directs them inside the ear.

overfishing The practice of fishing to such a degree that the fish population is used up.

paramecium (plural **paramecia**) A protist that has cilia and two kinds of nuclei.

pattern A consistent arrangement of shapes or colors.

perennial Lasting through the year or many years.

period of revolution The time it takes one body to revolve around another body.

period of rotation The time it takes a body to rotate once on its axis.

permafrost An underground layer of frozen soil that stays frozen for two or more years.

phases of matter The forms in which matter exists. The three major phases of matter are solid, liquid, and gas.

phases of the Moon The apparent changes in size and shape of the side of the Moon facing Earth that is lighted by the Sun.

phenotype The observable characteristics of an organism that are determined by genotype; the expression of specific traits.

photosphere The outermost layer of the Sun, which is around the convection zone and is actually the first layer of the Sun's atmosphere.

photosynthesis The process by which green plants use chemicals (water and carbon dioxide) in the presence of chlorophyll and light to produce food.

phototropism Growth or movement of a plant in response to light.

physical properties Characteristics of matter that can be measured and observed without changing the makeup of the substance.

physical reaction A change in which no new substances form.

pigment A substance that gives color to materials.

plant hormones Chemicals in plants that control cell growth.

plumule The tiny, immature leaves located at the tip of an epicotyl that at maturity form the first true leaves of a plant.

polymer A very long chain-like molecule.

population All the organisms that occur in a specific habitat or that are the same kind or species.

positive gravitropism Downward growth or movement of a plant, in the direction of the force of gravity.

positive phototropism   Growth or movement of a plant toward light.

potential energy (PE)   Stored energy of an object due to its position or condition.

pound   The English unit of weight.

power grip   A tight hold on a large object in which the fingers and thumb are wrapped around the object.

precipitation   Any form of water that falls from the atmosphere.

precision grip   A tight hold on a small object in which the fingers and thumb pick up the object.

predator   An animal that hunts other animals for food.

prey   An animal that a predator feeds on.

primate   A mammal that has grasping hands, including humans, apes, and monkeys.

protective coloration   Body coloration or a pattern that helps to camouflage animals from predators.

protist   An organism of the kingdom Protista, which includes most of the unicellular organisms having visible nuclei.

proton   A positively charged particle in the nucleus of an atom.

Punnett square   A grid used to determine the percentage of possible genotypes of offspring based on parental gene combinations.

pure trait   An offspring whose alleles are the same for a trait.

radiant energy   A form of energy that travels in waves.

radiation   Radiant energy; also the transmission of radiant energy in waves.

radiation zone   The layer of the Sun between the core and the convection zone.

radicle   The lower part of a hypocotyl that develops into a plant's roots.

rain forest   See **tropical forest.**

reaction time   The time it takes an organism to respond to a stimulus.

recessive allele   A gene form that does not determine a trait when a dominant allele is present.

reproduction   The process by which an organism produces young of the same species.

reradiate   To emit previously absorbed radiation.

resolving power   The eye's ability to focus on objects at a distance.

response   A reaction of an organism to a stimulus.

revolve   To move in a circular path around another object.

rift valley   A deep, narrow crack in Earth's crust along the top of a midocean ridge.

rock   A solid mixture of usually two or more minerals.

root system   The part of a plant that grows down into the soil, from which it takes in water and nutrients.

rotate   To turn on an axis.

seafloor spreading   The process by which new oceanic crust is created and moves slowly away from the midocean ridges.

season   A regularly recurring period of the year characterized by a specific type of weather.

sedimentary rock   Rock formed of sediments that are deposited by water, wind, or ice.

sediments   Fragments of a material that have been carried from one place and deposited in another by an agent of erosion.

seed   A product of sexual reproduction in plants that contains genetic material from both parents and can develop into a mature plant.

seed coat   The protective outer covering of a seed.

seed cone   A cone that contains seeds.

semiarid   Describing a dry climate of low rainfall, but not as dry as a desert.

sex cells   Specialized cells, sperm and eggs, produced by meiosis.

sex chromosome   A chromosome that contains the gene for gender and is known as an X or Y chromosome.

sexual reproduction   Reproduction by fertilization.

shore   The land at a shoreline.

shoreline   A border where a body of water meets the land.

sidereal day   The time it takes Earth to make one complete rotation of 360°.

silicates   Chemicals in Earth's mantle that consist of the elements silicon and oxygen combined with another element.

solar eclipse   An eclipse in which the Moon passes in front of and blocks the light of the Sun.

solid   A substance in a phase of matter characterized by a definite shape and volume.

solution   A homogeneous mixture in which one substance is dissolved in another.

sound   Energy that moves as waves through air or other materials.

Southern Hemisphere   The region below, or south of, Earth's equator.

South Pole   The south end of Earth's axis.

species   A group of similar organisms that can produce more of their own kind.

speed   The rate at which a distance is traveled in a given time.

sperm   A male sex cell.

spiderling   A young spider.

standard   A material to which other materials are compared.

static electricity   A buildup of electric charges, either positive or negative.

stem   The central support structure of a plant.

stimulus   (plural **stimuli**) Something that causes a response in an organism.

stratosphere   The layer of Earth's atmosphere between the mesosphere and the troposphere.

substance   A material that is made of one kind of material.

summer solstice   The day when Earth's North Pole is tilted closest to the Sun, on or about June 21 in the Northern Hemisphere.

surface gravity   Gravity at or near the surface of a celestial body.

tectonic plates   Rigid pieces of the lithosphere that cover Earth's surface.

temperate forest   A forest in a temperate zone; also called **deciduous forest.**

temperate zone   Either of two regions between latitudes 23.5° and 66.5° north and south of the equator.

temperature   The physical property that determines the direction that heat flows between substances.

thermal energy   The total internal energy of a material due to molecular motion.

thermometer   An instrument used to measure temperature.

thermosphere   The layer between Earth's atmosphere and the exosphere.

third quarter moon   The phase of the Moon that follows the full moon in which half of the side of the Moon facing Earth is lighted.

total solar eclipse   A solar eclipse in which the Moon blocks all the Sun's light.

trait   A physical characteristic.

tree line   A border between a region with trees and a tundra.

tropical forest   A forest in the tropical zone that is made up of tall broad-leaved evergreen trees that form a canopy; also called **rain forest.**

tropical zone   The region of Earth between latitudes 23.5°N and 23.5°S that is characterized by hot, humid, wet weather throughout the year.

tropism   The bending movement of a plant in response to a stimulus, such as light, heat, water, or gravity.

troposphere   The lowest, innermost layer of Earth's atmosphere.

tundra   A treeless biome, mostly of the Arctic region.

unbalanced force   Force applied on an object without an opposing force of equal strength.

unicellular   One-celled.

vegetative propagation   Production of a new plant from a plant part other than a seed.

veins   Large blood vessels that carry blue oxygen-poor blood to the heart.

venules   Ends of veins that connect to capillaries.

volume   The amount of space something occupies.

wane   To get smaller.

warm-blooded animal   An animal that generates heat to maintain a constant internal body temperature.

water cycle   The cycle of evaporation and condensation that moves Earth's water from one place to another.

water vapor   Water in the gas phase.

water wave   A disturbance on the surface of water that repeats itself.

waves   Disturbances that move through matter or space.

wax   To get larger.

weathering   The stage of erosion that involves only the breakdown of crustal materials.

weight   A measure of the force of gravity, which on Earth is a measure of the force with which Earth's surface gravity pulls on an object.

winter solstice   The day when Earth's South Pole is tilted farthest away from the Sun, on or about December 22 in the Northern Hemisphere.

work (w)   The movement of an object by a force.

zygote   A cell formed by the joining of a sperm and an egg.

Index

abiotic, 106

absorb, 24, 25, 52, 64, 65, 135, 136, 194

accelerate, 35

active layer, 107, 108, 194

adaptation, 129, 130, 131, 194

agent of erosion, 148, 149, 194

air mass, 170, 171, 194

allele, 73, 74, 194

Alpha Centauri, 173

alpine tundra, 107, 109, 110, 194

amphibians, 193

anatomy, 55

angular distance:
  definition of, 181, 182, 194
  hand measurements, 181, 182

annular eclipse, 187

Antarctica, 121, 122

anthocyanin, 64

aphelion, 179

apogee, 183

apparent size, 187, 189, 194

Arctic Circle, 109

Arctic region, 107, 108, 194

Arctic tundra, 107, 108, 194

Aristotle, 58

arteries, 66, 67, 194

arterioles, 66, 67, 194

asexual reproduction, 88, 89, 194

asthenosphere, 146, 158

astronomy, 143, 172–188

atmosphere:
  definition of, 145, 166, 167, 194
  Earth's, 145, 166, 167
  Sun's, 173

atom, 12, 13, 194

attraction, 24, 25, 194

auxin, 99, 100, 194

average speed, 27, 28, 194

axis, 175, 176, 194

balanced force, 35, 36, 194

ballooning, 97, 98, 194

banyon tree, 103

beach, 162, 163, 195

behavior, 90–105
  definition of, 91, 92, 194

biome:
  definition of, 106, 107, 108, 194
  desert, 121, 122

birds, 93, 133, 139

bisexual, 109

biosphere, 145

biotic, 166

bison, 118

Bjerknes, Vilhelm, 170

blood, 66, 67, 194

blue moon, 185

bond, 12, 13, 194

boreal forest, 109

breccia, 150

bristlecone pine, 109

butterfly, 141, 142

bulb, 88

caloric theory, 46

calving, 164

camouflage:
  definition of, 125, 126, 194
  protective coloration, 130, 131

canine, 135

canopy, 115, 116, 194

capillaries, 66, 67, 194

capillary link, 66, 67, 194

carotene, 64

celestial bodies, 31, 143, 194

cell, 62, 63, 194

cell membrane, 62, 63, 194

Celsius, Anders, 48

Celsius scale, 48, 49, 194

cementation, 150, 151, 194

center of gravity, 26

center of mass, 26

cerebellum, 95, 96, 194

cerebrum, 95, 96, 195

characteristics, 58, 59, 195

chemical property, 18, 19, 195

chemical reaction, 18, 19, 195

chemical sedimentary rock, 150, 156

chemistry, 9

chemical weathering, 148, 149, 159, 195

chlorophyll, 64, 65, 195

chromatography, 24, 25, 195

chromosome, 73, 74, 195

chromosphere, 173

cilia, 68, 69, 195

circulatory system, 66, 67, 195

classification, 57
  definition of, 58, 59, 195
  Linnaeus, Carolus, 57, 58

clastic rock:
  definition of, 150, 151, 195
  types of, 150

clastic sedimentary rock, 150

clone, 88

cold-blooded animal, 93, 94, 195

cold desert, 122, 123, 195

colloid, 24

community, 106

compaction, 150, 151, 195

compound, 12, 13, 195

conclusion, 7

condensation, 160, 161, 195

conduction:
  definition of, 46, 47, 195
  investigation of, 47, 53

cone, 109, 110, 195

conglomerate, 150

conifer, 109, 110, 195

coniferous forest, 109, 110, 195

conservation of energy, 41

conservation of matter, 12, 18

contract, 20, 21, 195

convection, 50, 51, 195

convection currents, 50, 51, 195

convection zone, 173, 174, 195

convergent boundary, 158

core:
  definition of, 141, 142, 173, 174, 196
  Earth's, 146, 147
  Sun's, 173, 174

corona, 173

cotyledon, 86, 87, 195

country rock, 148, 149, 195

crescent moon, 185, 186, 195

cross-linked, 18, 19, 195

cross-staff, 181

crust, 146, 147, 195

crystal, 156, 157, 195

currents, 50

cutting, 88, 89, 195

cytoplasm, 62, 63, 195

cytosol, 62

decelerate, 35

deciduous, 111, 112, 195

deciduous forest, 111, 112, 195

degree, 181, 182, 195

density, 14, 164, 165, 195

deoxyribonucleic acid (DNA), 73, 75, 195

desert, 121, 122, 195

diatomic molecule, 12, 13, 195

diffuse, 16, 17 195

dissolve, 22, 23, 195

divergent boundaries, 158, 159, 195

diversity, 129

dominant allele, 81, 83, 195

drag line, 97

due north, 33

eagle, 139

earth science, 143–171

eclipse:
  annular, 187
  definition of, 187, 188, 196
  solar, 187, 188

ecosystem, 106–128
  definition of, 106, 107, 108, 196

ectotherm, 93

egg, 77, 78, 196

elasticity, 20, 21, 196

electromagnetic waves, 52, 53, 196

electron, 11, 12, 37, 38, 196

element, 12, 13, 196

elliptical, 187, 188, 196

embryo, 86, 87, 196

endangered, 127

endotherm, 93

energy (E), 41–54
  conservation of, 41, 44–45
  definition of, 42, 43, 196
  forms of, 41, 42
  groups, 41, 42
  heat, 52

environment, 91,106, 107, 108, 196

epicotyl, 86, 87, 196

equator, 111, 112, 196

erosion, 148, 149, 196

ethologist, 91

ethology, 90

evaporate, 16, 17, 196

evaporation, 93, 156, 160, 161, 196

evaporite, 150, 156

evergreen, 109, 110, 196

experiment, 7

exosphere, 166, 167, 196

expand, 46, 47, 196

extinct, 127, 128, 196

Fahrenheit, Gabriel, 48

Fahrenheit scale, 48

fertilization, 81, 82, 196

first quarter moon, 185, 186, 196

fluid, 16, 17, 24, 196

force, 26–40
  definition of, 27, 28, 196

forest, 109, 110, 196

formula, 12, 13, 196

fossil fuels, 168

Franklin, Benjamin, 50

friction, 26, 39, 40, 196

front:
  cold, 171
  definition of, 170, 171, 196
  occluded, 171
  warm, 171

full moon, 185, 186, 196

gas, 16, 17, 196

gender, 77, 78, 196

gene, 73, 75, 196

gene site, 73, 75, 196

genotype, 81, 83, 196

geosphere, 145, 146, 147, 158, 196

geotropism, 99, 103, 104, 196

germination, 103, 104, 196

gibbous moon, 185, 186, 196

glacier, 164, 165, 196

glide, 141, 142, 196

graduated cylinder:
  definition of, 14, 15, 196
  model of, 15, 189–190

grassland, 118, 119, 196

gravitational, 26

gravitational potential energy (GPE), 42, 43, 196

gravitropism, 103, 104, 196

gravity, 29, 30, 196

gravity rate (G.R.), 31, 32, 196

greenhouse effect, 168, 169, 196

grip, 133, 134, 196

Gulf Stream, 50

habitat, 106

halite, 156, 157, 197

hand-eye coordination, 95, 96, 197

headland, 162

heat, 46, 47, 53, 196

heredity, 70–89
  definition of, 71, 72, 197

heterogeneous, 22, 23, 197

homogeneous, 22, 23, 197

Hooke, Robert, 62

horizon, 177, 178, 197

humidity, 115, 116, 197

hybrid, 84, 85, 197

hydrosphere, 145

hydrotropism, 99

hypocotyl, 86, 87, 197

hypothesis, 7

iceberg:
  calving, 164
  definition of, 164, 165, 197
  tabular, 164
  tongue of, 164

igneous rock, 152, 153, 197

inborn, 91

incandescent, 52

inertia, 39, 40, 197

infrared radiation, 53, 54, 197

infrared rays, 52, 53, 197

inherit, 71, 72, 197

innate behavior, 91, 92, 93, 197

inorganic, 148

instinct, 141

insulator:
  atmosphere, 166
  definition of, 46, 130, 131, 197

International System of Units (SI), 14, 15, 197

ionosphere, 166

jade, 154

joule (J), 42, 43, 197

kinetic energy (KE), 42, 43, 197

kinetic mechanical energy. *See* kinetic energy

latitude, 111, 112, 197

lava, 152, 158, 159, 197

Lavoisier, Antoine, 46

law of conservation of energy, 41, 44, 45, 197

law of conservation of matter, 41, 44, 45, 197
leap year, 175
learned behavior, 91, 92, 93, 197
Le'on, Ponce de, 50
life science, 55–142
  anatomy, 55
  botany, 55
  definition of, 55
  zoology, 55
lift, 141, 142, 197
light-year, 173
Linnaeus, Carolus, 57, 58, 59
lion, 135, 136
liquid, 16, 17, 197
liter (L), 14, 15, 197
lithification, 150, 151, 197
lithosphere, 158, 159, 197
living systems, 57
locomotion, 68, 69, 197
lunar synodic month, 185
Lunik 3, 183

macromolecules, 18
maglev, 35
magma, 152, 153, 197
magnet, 33, 34, 197
magnetic field, 33, 34, 197
magnetic force, 26, 33, 34, 197
magnetic lines of force, 33, 34, 197
magnetic materials, 33, 34, 197
magnetic poles, 33, 34, 197
magnetism, 33, 34, 197
magnetosphere, 33, 166
mammal, 93, 133, 134, 197
mantle, 146, 147, 197
mass, 12, 13, 14, 197
matter, 11–25
  conservation of, 41
  definition of, 12, 13, 197
  pull of gravity on, 31
mechanical energy:
  conservation of, 44, 45
  definition of, 44, 45, 197
mechanical weathering, 148, 149, 197

meiosis, 77, 78, 197
melt, 152, 153, 198
melting point, 152, 153, 198
meniscus, 14, 15, 198
mesosphere, 166, 167, 198
metamorphic rock, 154, 155, 198
metamorphism, 154, 155, 198
metric system, 14, 15, 198
microwave, 52
midocean ridge, 158, 159, 198
migration, 141, 142, 198
milliliter (ml), 14, 15, 198
mineral, 148, 149, 198
mixture, 22, 23, 198
molecule:
  definition of, 12, 13, 198
  diatomic, 13
  models of, 13
  polar, 52
molten, 152, 153, 198
monomer, 18
motion, 27, 28, 198
multicellular, 57
muscular force, 27, 28, 198
mutualism, 130

negative gravitropism, 103, 104, 196
negative tropism, 101
neutron, 12
new moon, 185, 186, 198
newton (N), 31, 32, 198
Niagara Falls, 148
niche, 118
nictitating membrane, 139
non-Newtonian fluid, 18, 19, 198
North American Plate, 158
North Pole, 177, 178, 179, 198
north pole of a magnet, 33
Northern Hemisphere, 177, 178, 198
nuclear fusion, 173, 174, 198
nucleus (plural nuclei):
  definition of, 12, 37, 38, 62, 63, 198

orbit, 175, 176, 198
organic sedimentary rock, 150
organisms, 57, 58, 59, 198
outer ear, 137, 138, 198
overfishing, 127, 128, 198
ozone, 166

panda, 133
paramecium, 57, 68, 69, 198
parasite, 130
parasitism, 130
parrot, 133
partial solar eclipse, 187
Pasteur, Louis, 68
pattern, 125, 126, 198
penguin, 121, 122
penumbra, 187
perennial, 107
perennial plant, 107
perigee, 183
perihelion, 179
period of revolution, 183, 184, 198
period of rotation, 183, 184, 198
permafrost, 107, 108, 198
phases of matter, 11, 16, 17, 198
phases of the moon, 185, 186, 198
phenotype, 71, 72, 81, 83, 198
photosphere, 173, 174, 187, 198
photosynthesis, 64, 65, 198
phototropism, 99, 101, 102, 198
physical properties, 11, 14, 16, 17, 198
physical reaction, 16, 17, 198
physical science, 9–54
  definition of, 9
  energy, 41–54
  forces and motion, 26–40
  matter, 11–25
physics, 7
pigment:
  anthocyanin, 64

carotene, 64
chlorophyll, 64
  definition of, 24, 25, 198
plant hormones, 99, 100, 198
plasma, 66
plate tectonics, 145
plumule, 86, 87, 198
polar bear, 130
polar molecule, 52
pollen cone, 109
polymer, 18, 19, 198
population, 106, 121, 122, 198
positive gravitropism, 103, 104, 198
positive phototropism, 101, 102, 198
positive tropism, 101
potential energy (PE), 42, 43, 199
potential mechanical energy. See potential energy
pound, 31, 32, 199
power grip, 133, 134, 199
precipitation, 160, 161, 199
precision grip, 133, 134, 199
predator, 123, 124, 199
prey, 123, 124, 199
primates, 133, 134, 199
problem, 7
protective coloration, 125, 126, 199
protist, 68, 69, 199
protoctista, 68
proton, 12, 37, 38, 199
Punnet square, 84, 85, 199
pure trait, 84, 85, 199

radial sesamoid, 133
radiant energy:
  definition of, 52, 53, 199
  solar, 168, 173
radiate, 52
radiation:
  definition of, 52, 53, 199
  forms of, 52
  heat, 52
  infrared, 52, 53, 197

radiation (continued)
  properties of, 52, 53, 137
  solar, 52
  X ray, 52
radiation zone, 173, 174, 199
radicle, 86, 87, 199
reaction time, 95, 96, 199
recessive allele, 81, 83, 199
reflected, 24, 64
reindeer, 131
reproduction, 70–89, 199
reptile, 93
reradiate, 168, 169, 199
research, 7
resolving power, 139, 140, 199
response, 91, 92, 199
revolve, 175, 176, 199
rift valley, 158, 159, 199
rock, 148, 149, 199
rock cycle, 145, 154
rock salt, 156
root system, 86, 87, 199
rotate, 175, 176, 199
rubber, 18, 20
Rutherford, Count, 46

salivate, 91
salt water, 164
sandstone, 150
scientific method, 7–8
seafloor spreading, 158, 159, 199
season:
  cause of, 179
  definition of, 111, 112, 199
sedimentary rock, 150, 151, 199
sediments, 148, 149, 199
seed, 86, 87, 199
seed coat, 86, 87, 199
seed cone, 109, 110, 199

semiarid, 118, 119, 199
sex cells, 77, 78, 199
sex chromosome, 77, 78, 199
sexual reproduction, 81, 82, 199
shale, 150
shadow, 187
shore, 162, 163, 199
shoreline, 162, 163, 199
sidereal day, 175, 176, 199
silicates, 146, 147, 199
solar day, 175
solar eclipse, 187, 188, 199
solar energy, 156
solar radiation, 52
solid, 16, 17, 199
solute, 22
solution:
  definition of, 22, 23, 200
  types, 22
solvent, 22
sonar, 145
sound, 137, 138, 200
Southern Hemisphere, 177, 178, 200
South Pole, 177, 178, 179, 200
south pole of a magnet, 33
species, 129, 130, 131, 200
speed, 27, 28, 200
sperm, 77, 78, 200
spiderling, 97, 98, 200
standard, 20, 21, 200
static, 26
static electricity, 37, 38, 200
stem, 86, 87, 200
stimulus, 90, 91, 92, 200
stratosphere, 166, 167, 200
substance, 12, 22, 200
summer solstice, 179, 180, 200
Sun, 52

solar radiation, 52
surface gravity, 31, 32, 200
sweat, 93
symbiotic relationship, 130
system, 12, 57
systematic classification, 58

table salt, 156
taiga, 109
taxonomy, 58
tectonic plates, 158, 159, 200
temperate forest, 111, 112, 200
temperate zones, 111, 112, 200
temperature, 48, 49, 200
thermal energy, 46, 47, 200
thermometer:
  Celsius scale, 48, 49, 191–192, 194
  definition of, 48, 49, 200
  Fahrenheit scale, 48
  types of, 48, 49
thermosphere, 166, 167, 200
thigmotropism, 99, 101
timberline, 109
third quarter moon, 185, 186, 200
tongue, 164
torque, 26
total solar eclipse, 187, 188, 200
trait, 71, 72, 200
transform boundary, 158
transmitted, 24, 64
transpiration, 160
tree line, 109, 110, 200
trees, 111
tropical forest, 115, 116, 200

tropical zone, 115, 116, 200
tropism, 99, 100, 200
troposphere, 166, 167, 200
true north, 33
tuber, 88
tundra, 107, 108, 200
turgor pressure, 99

umbra, 187
unbalanced force, 35, 36, 194
unicellular, 68, 69, 200

vegetative propagation, 88, 89, 200
veins, 66, 67, 200
venule, 66, 67, 200
volume, 14, 15, 200

wane, 185, 186, 200
warm–blooded animal, 93, 94, 133, 200
waste, 66
water cycle, 160, 161, 200
water vapor, 160, 161, 200
water wave, 162, 163, 200
wave:
  definition of , 52, 53, 200
  electromagnetic, 52, 53
wax, 185, 186, 200
weather, 166
weathering, 148, 149, 200
weight, 14, 31, 32, 200
weightlessness, 29
winter solstice, 179, 180, 200
work, 42, 43, 200

xenolith, 152
X ray, 52

zoology, 55
zygote, 81, 83 200